4 PHASE APPROACH TO COMPETITIVE ADVANTAGE

Creating Organizational Competitive Advantage

ROBERT A. CARTIA

Many thanks to those who have been encouraging me to get my ideas into print so the world could benefit. Special thanks to the late Tom Saaty (1926 - 2017) who, after our last conversation, altered my life trajectory; my uncle Sam who once told me to never be that man who has let life pass him by wishing he had taken advantage of what life had to offer; my mother Linda who has always believed in me; and my wife Kathy and children Merissa and Giavonna who support my dreams.

"In business as it is in life, relevance is everything. Your objective is to remain relevant. Always be transforming."

Robert A. Cartia (2017)

FOREWORD

I recall first meeting and working with Rob nearly twenty years ago while working in the metals industry. He introduced me to a notebook he always carried with him that he dubbed as his "Book of Knowledge". This was his earliest manuscript of this book you're holding.

Rob has been developing this book over his entire career. It's the product of his experiences as a leader in operations and operational excellence. He excels at strategy development and strategy deployment to achieve competitive advantage and continuous improvement. It's based not only in abstract principles but also concrete experiences with proven results.

Chris Gill, P.E.- Engineering Manager, Georgia Pacific

Table of Contents

INTRODUCTION

The intent of this book is to provide real world applications and proof point results. As compared to the many published business transformation books out there, this book goes beyond simply stating theory. Other business books often leave the reader hanging as to whether the stated content actually works. However, the content provided in this book is based on experience and trial and error. It is a proven blueprint.

Some of my views and approaches represent hybrid thinking which has worked for me as a result of situations I have faced. The 4 Phase Approach to Competitive Advantage Model is a direct result of my experience gained over the past 28 years working within eight companies spanning 10 industries. **(Figure I-1)**

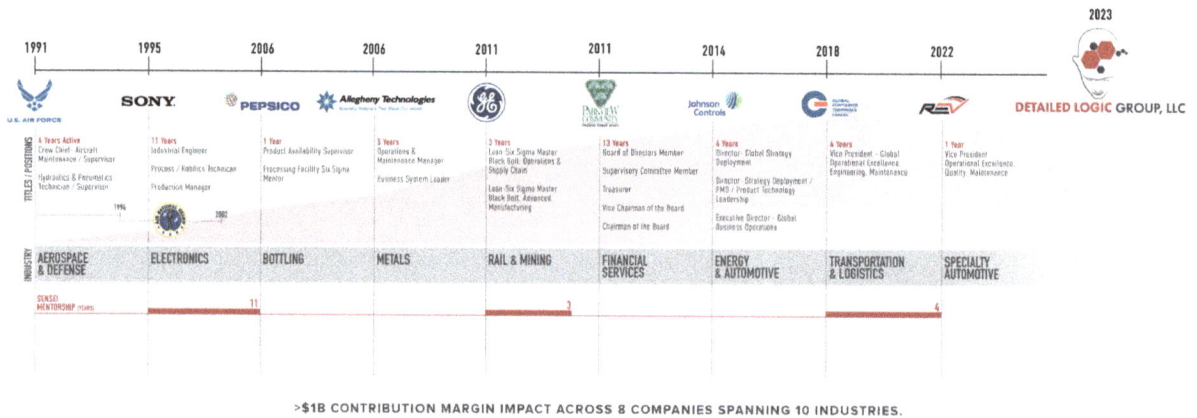

>$1B CONTRIBUTION MARGIN IMPACT ACROSS 8 COMPANIES SPANNING 10 INDUSTRIES.

I officially made public the framework of The 4 Phase Approach Model in October of 2017 by publishing it on one of my social media pages. Two years later in March of 2019, I presented the framework in Orlando at the 2019 Business Transformation & Operational Excellence Summit (BTOES) World Summit & Industry Awards Event. The overwhelming response in 2017 and again in 2019 was that I share the detail of the Model, so I began writing this book.

I have created a proven approach that if done correctly, will result in competitive advantage for any company in any industry. **(Figure I-2 & I-2b)**

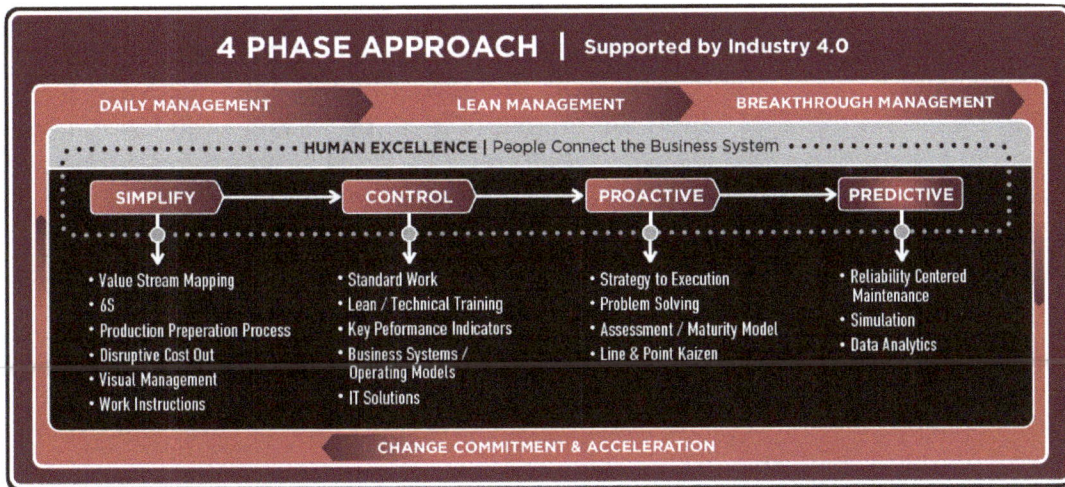

4 PHASE APPROACH | Supported by Industry 4.0

DAILY MANAGEMENT	LEAN MANAGEMENT	BREAKTHROUGH MANAGEMENT

HUMAN EXCELLENCE | People Connect the Business System

SIMPLIFY	CONTROL	PROACTIVE	PREDICTIVE

• Value Stream Mapping
• 6S
• Production Preperation Process
• Disruptive Cost Out
• Visual Management
• Work Instructions
	• Standard Work
	• Lean / Technical Training
	• Key Peformance Indicators
	• Business Systems /
	  Operating Models
	• IT Solutions
		• Strategy to Execution
		• Problem Solving
		• Assessment / Maturity Model
		• Line & Point Kaizen
			• Reliability Centered
			  Maintenance
			• Simulation
			• Data Analytics

CHANGE COMMITMENT & ACCELERATION

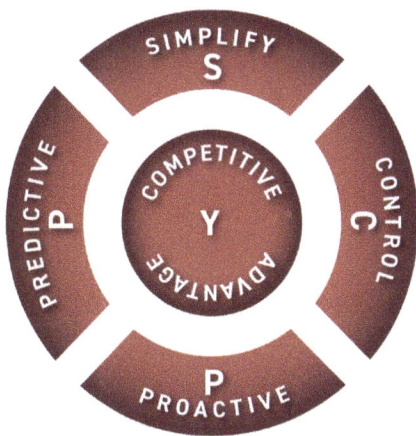

$$Y = f (X_1, X_2, X_3, X_4)$$

$$Y = (S, C, P, P)$$

COMPETITIVE ADVANTAGE   SIMPLIFY   CONTROL   PROACTIVE   PREDICTIVE

The Model is comprised of:

- 4 Phases- The pillars of the Model.

- Tools Per Phase- The tools used to execute each phase of the Model.

- Execution Management System (Daily, Lean, and Breakthrough Management)- The vehicle that drives the Phases of the Model.

- Human Excellence- People connect the business system through application of the Model.

- Change Commitment & Acceleration- Drives commitment vs. compliance with the change that is required as a result of the Model.

- Industry 4.0- Technologies and solutions that support the Model.

This 4 Phase Model applies to manufacturing, services, and all functions across any organization's value stream. The tools that are provided can be added or removed based on the application. As an example, if the Model were being used to drive Procurement excellence, you would remove tools such as: Production Preparation Process and Reliability Centered Maintenance.

This method has enabled me, working with many great teams and organizations, to generate over one billion dollars in contribution margin over my 28 year career. These results highlight the "what is possible" where today's breakthroughs become tomorrow's daily management. Doubling revenue and profit within a specific time frame can only be accomplished through this iterative approach. **(Figure I-3)**

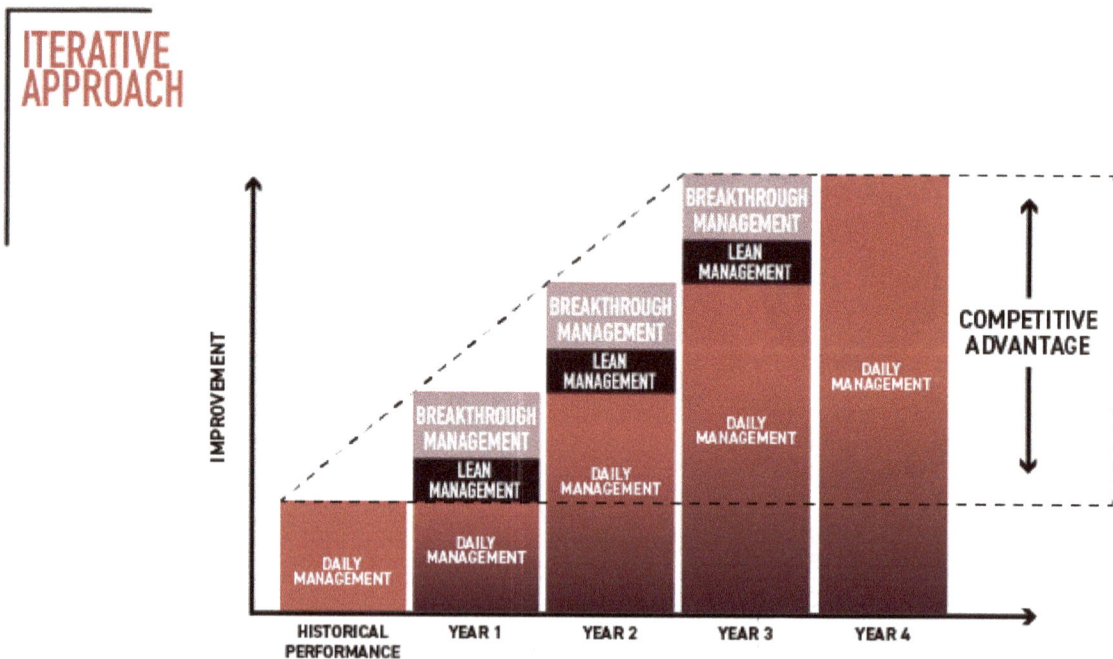

This model serves as a playbook for how an organization should manage its time and one that is focused on value generation and the acceleration of innovation. Daily Management, Lean Management, and Breakthrough

Management are components of the Execution Management System which functions as the vehicle to drive the Model Phases. **(Figure I-4)**

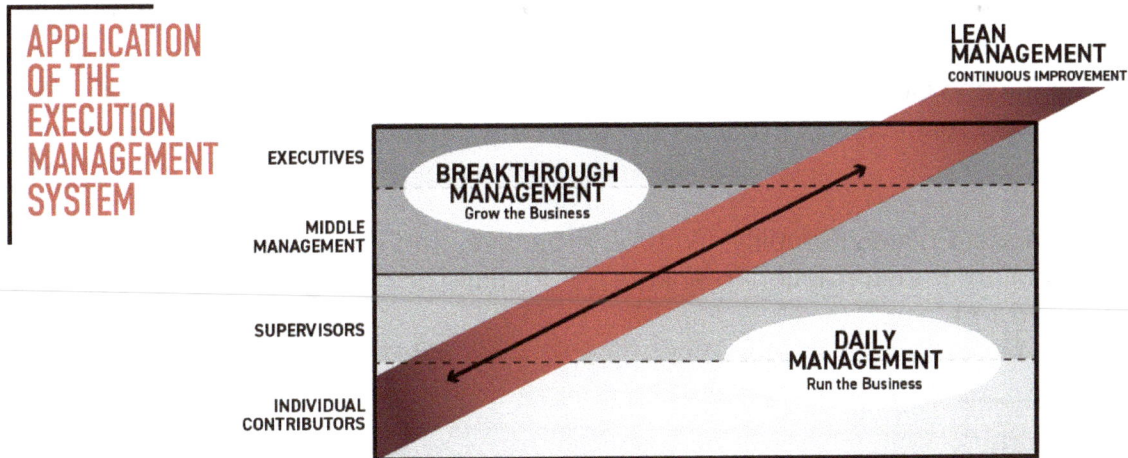

APPLICATION OF THE EXECUTION MANAGEMENT SYSTEM

LEAN MANAGEMENT
CONTINUOUS IMPROVEMENT

EXECUTIVES

BREAKTHROUGH MANAGEMENT
Grow the Business

MIDDLE MANAGEMENT

SUPERVISORS

DAILY MANAGEMENT
Run the Business

INDIVIDUAL CONTRIBUTORS

To put this system into perspective, let me use an example of how it applies to our personal lives. Daily Management is the operating rhythm cadence; about our daily routines. As part of my routine, I do three days of cardio (treadmill). Lean Management is about correcting a problem or improving some aspect within Daily Management. My problem is that I am not losing as much weight as I am targeting so I increase the duration of time I do cardio for each of the three days. Breakthrough Management is about engaging something new: a "do different." It is used when no more improvement can be achieved through Lean Management, or when innovation is required to achieve the targeted results. When I reach a plateau with my cardio, I decide to replace two of the three days of cardio with high intensity mixed martial arts. Through this example of the Execution Management System, I am able to achieve (or exceed) my desired weight loss goals.

The Model was developed through the synthesis of my experience with Lean and Reliability Centered Maintenance (RCM). The Model's four steps are:

1. **Simplify:** The first step is focused on how companies define the way they make decisions; develop and commonize / standardize processes; work together with internal staff; and work with their partners and customers.

2. **Control:** The second step is focused on sustaining results through control mechanisms and organizational discipline.

3. **Proactive:** The third step is focused on initiating actions to continuously advance and innovate thus assisting the organization.

4. **Predictive:** The final step is focused on leveraging data analytics to anticipate and react to problems before they occur thus informing the organization.

Simplify and Control enables the organization to disrupt the current game whereas Proactive and Predictive enables the organization to dictate the future game.

The Model Phases (Simplify, Control, Proactive, and Predictive) become states in constant transformation to ensure competitive advantage is maintained. **(Figure I-5)**

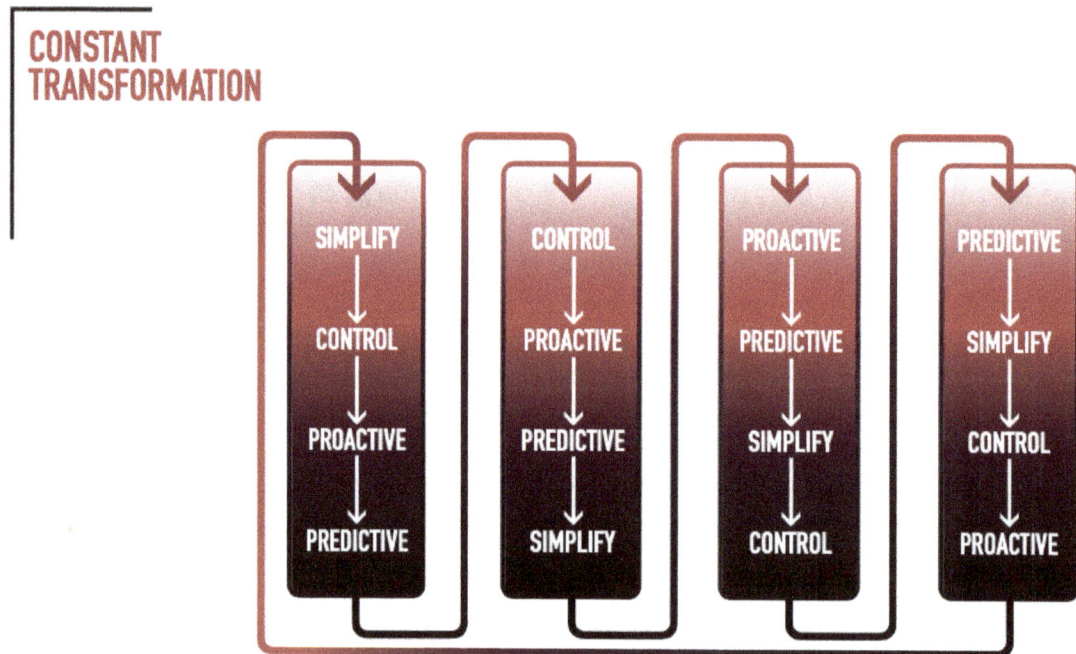

CONSTANT TRANSFORMATION

SIMPLIFY	CONTROL	PROACTIVE	PREDICTIVE
CONTROL	PROACTIVE	PREDICTIVE	SIMPLIFY
PROACTIVE	PREDICTIVE	SIMPLIFY	CONTROL
PREDICTIVE	SIMPLIFY	CONTROL	PROACTIVE

The 4 Phases will provide an organization the ability to achieve "true north" growth quicker. In the past where it took a company decades to double profit, transformative growth can now be achieved within a five year strategic plan cycle. **(Figure I-6)**

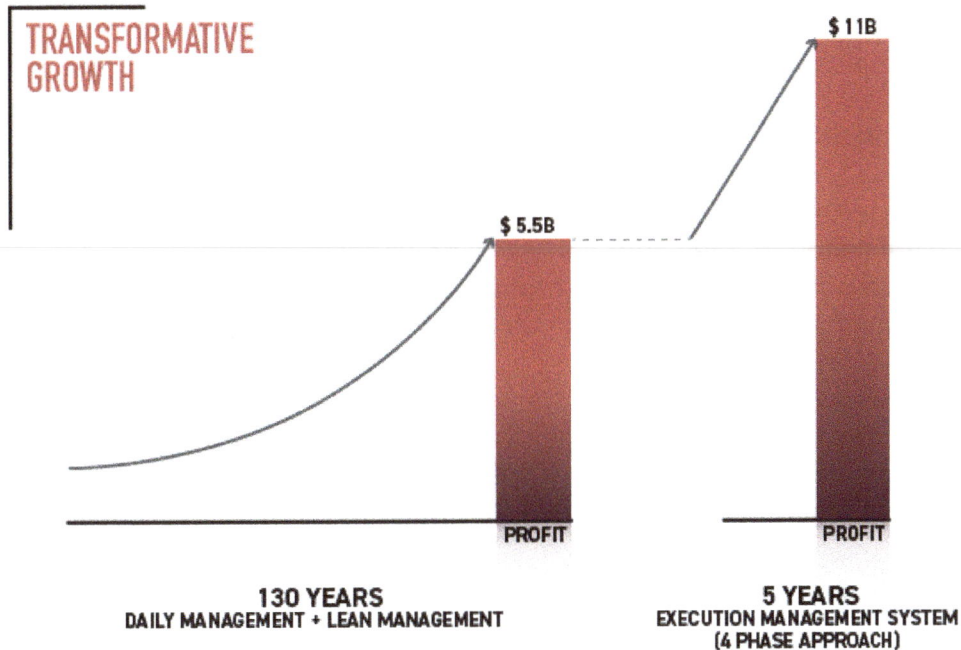

Independently, these Model Phases are not new. But codified and synchronized, they become a new way of thinking; a structure to advance processes; an improvement of the financial health of organizations. Together, these actions place an organization in a more desirable position with respect to competitive advantage.

In Chapter 1, I will detail how people make the model work. In Chapter 2, I will explain the vehicle that enables this approach. In Chapter 3, I will discuss competitive advantage and the most over-looked forms of it. In Chapter 4, I will detail the experience factor that led to this approach. In Chapter 5 through 8, I will walk through each of the 4 Phases to include Use Cases (Application Benefits) and Caution Points- Where Companies Can Get This Wrong. In Chapter 9, I will outline how Industry 4.0 technologies support The 4 Phase Approach Model. In Chapter 10, I will pull it all together. Within each Phase, I will walk through specific Lean tools. Even though a specific tool may be assigned to a specific Phase does not necessarily mean that it cannot be used within an additional Phase or Phases. For purposes of this book, I have assigned each Phase to situations where real world examples have shown the biggest impact. In addition, please note that Industry 4.0 Technologies can be used across each Phase as a means to support improvements.

Organizational complacency is a killer and thus you can quickly become your own worst enemy. Maintaining an aggressive approach toward your competitors is a must to not only disrupt the game, but to dictate the game. The wins do not last long before you need to transform the organization once again. Just when an organization achieves a desired state of optimization through the transformation, it may be time to transform once again.

As a prelude to the coming chapters, I want to establish a few points that will provide clarification to a few terms that will be used:

1. Business Transformation is when there is more than one discrete change or area of improvement occurring within an organization. As an example, if a business system, an operating model, and a training program were being enhanced or rolled out as part of a blueprint, this would be considered Business Transformation. If there was just one discrete area of improvement or change, that would be referred to as Change Management.

2. Operational Excellence is the end result of organizational improvement; it is a result that is achieved.

3. Continuous Improvement is the journey towards achieving Operational Excellence.

4. Lean is simply a mindset; a way of thinking. It is also a practice of continuous improvement. It begins with thinking and ends with the completed continuous improvement actions required to achieve value (i.e., Operational Excellence).

5. Lean Tools are contained within Daily Management, Lean Management, and Breakthrough Management, and provide a means to achieve Operational Excellence.

The above terms may be used in different manners throughout different organizations. As an example, over the course of my career, I have worked within a Business System function, a Business Transformation function, and led Business Operations and Operational Excellence functions. They all had the objective of organizational improvement; driving cost leadership while having respect for people.

Organizations that leverage the benefits of having and operating with a Lean mindset are organizations that prosper. This is especially true during times of global / national uncertainty such as the 2008 Stock Market Crash and the 2020 COVID-19 Global Pandemic. During these times when the top line (revenue) of the financial income statement may not have increased, intelligent organizations leveraged the Lean mindset to manage the middle of the income statement (variable expenses) thus improving the bottom line (profit). This, in itself, is a competitive advantage. Lean thinking places an organization in a mindset to challenge the current state. Transformation is "living" Lean whereas implementation is "doing" Lean. In order to drive value, you must first know where the opportunities are and this is what the application of Lean tools exposes.

Lean is about ruthless prioritization; it's the "move with a purpose". It enables growth. It isn't growth in and of itself. Lean can be viewed as common sense methods that improve a current state through waste reduction. It is about moving the final value to the customer or end user. Lean methods should always focus on quality and ensuring high standards. It is critical for ensuring repeatable and sustainable processes and results. It becomes a blueprint for creating a culture of not only problem solvers but problem preventers; this is what is required for long-term success. Lean implementation becomes the "how" by which an organization ensures its success.

The five steps of Lean implementation according to James P. Womack and Daniel T. Jones in their published book, *Lean Thinking(1)* are:

1. **Identify Value:** The critical starting point for lean thinking is value. Value can only be defined by the ultimate customer. And it's only meaningful when expressed in terms of a specific product (a good or a service, and often both at once) which meet the customer's needs at a specific price and at a specific time.

2. **Map the Value Stream:** The value stream is the set of all specific actions required to bring a specific product through the three critical management tasks of any business: the problem solving task, the information management task, and the physical transformation task. Map the current state where waste is identified and categorized and map a future state where the waste has been eliminated.

3. **Create Flow:** Make the remaining, value-creating steps flow smoothly. Focus on reducing lead-time by identifying and eliminating barriers.

4. **Establish Pull:** Let the customer pull the product from you as needed rather than pushing products, often unwanted, onto the customer.

5. **Seek Perfection:** There is no end to the process of reducing effort, time, space, cost, and mistakes while offering a product which is ever more nearly what the customer actually wants.

This book is focused on these factors (reducing effort, time, space, cost, and mistakes) but I want to reinforce the importance of the value of time within today and tomorrow's ever-growing need to operate faster.

According to the BTOES Report(2) 2021/2022 titled *The Global State of Operational Excellence- Critical Challenges & Future Trends* (this report is the most comprehensive state of Operational Excellence report ever conducted with input from across 86 industry sectors, 43 job functions, 11 senior-levels, all 10 major regions of the world, and organizations of all sizes and types from public, privately held, to government agencies and nonprofits), the average lifespan of a company in the S&P 500 is only 14 years. To stay relevant companies more than ever need to adapt to changing business environments, and continuously improve and innovate. The report additionally states that Operational Excellence needs to be viewed as a critical strategic weapon by the corporate-level execution leadership teams.

The reports summarize the three major challenges facing Operational Excellence programs:
1. Changing & improving company culture
2. Driving an end-to-end Business Transformation program
3. Sustaining Operational Excellence improvements

There is no substitute for the experience I gained haven started my career within a Japanese-based company where a focus on Continuous Improvement (Kaizen- Japanese word for Continuous Improvement; change for the better) was the expectation and the process to achieve it was never viewed as a burden or additional / extra work. The company did not advertise a business system. Nor did the company ever tout specific Lean principles. Instead, we lived by a saying... "Proud but never satisfied." This summed up our Lean mindset... the continuous improvement journey is never completed. I began as an assembly line worker and worked myself up into the management level. During my 11 years within this environment, I never viewed continuous improvement as anything extra but rather as just part of my standard work. Once the company announced they were transferring the volume and thus operations out of the state, every employee was hard at work updating their resumes and preparing for new opportunities elsewhere. During this time, it was very common for team members to ask me if I would assist them with updating their resumes. On several occasions, when asked "What should I put that I did during my employment here?" I would respond with focus areas such as: Kaizen events, Value Stream Mapping, 6S (5S + Safety), A3 Problem Solving, Standard Work, etc. The response would often be "Why would I put that... that was just part of my job responsibilities." This period of time really made me realize just how deep within the company DNA that continuous improvement was embedded. These foundational skills served as an excellent benchmark when employees were transitioning to other companies.

For me, continuous improvement is something that we do in every aspect of our lives. Yet, most people will resist it while in the workplace based on their perception that it equates to additional work for them leading to the "Why change now...we have always done it this way here" mentality. It is something that has been essential yet understated within the organization over the years; the "flavor of the day" effect. Employees don't necessarily understand the value of Lean principles.

An inflection point occurred in my career in 2008 while working within a Business System function within the metals industry. I was requested by my manager to provide an eight-hour Introduction to Lean course broke up into two 4-hour days to employees coming off the 3rd shift. I remember being a bit nervous, even though I knew them all, as this would be my first time providing this type of training within an environment where "we have been down this road before with different leaders" was the common perception. I felt very privileged to do so as both of my grandfathers retired from the metals industry each with 4 decades of service.The first four-hour session did not go well. Approximately 10 workers, each with decades of company / industry experience, and who had endured prior process improvement propaganda came into the training room after working the night shift. I noticed that I was losing them as I was delivering the presentation content. As I was driving home after the first day of providing the training, I was thinking about how I could reach them and salvage this important training. I knew it was important but they didn't know it was important. I reflected on my personal experience when I realized that Lean training and how it became "sticky" for me was different. I thought about how Lean just isn't this "crap" that we do while at work... it was so much more; it is all around us in life. Eureka! That was it. The second 4-hour day arrived and I had a new strategy. Once the team were in the training room, I opened by announcing the following: "Today we are going to do something different. I am not going to use any presentation slides during our last four hours together. The one thing I need all of you to remember is that Lean is about the customer and about identifying areas to reduce or eliminate the non-value added work (waste) within our company that the customer is not paying for." They were relieved and perhaps shocked that this is all I was asking them to remember from the 8-hour training session. They all wrote what I had said down in the event they were asked what they took away from the training. I followed this by asking them how many of them owned Harley Davidson Motorcycles. The majority of them said they did. Once they were done talking about their bikes, I asked them how many of them owned John Deere Tractors. Much like their response to my previous question, most of them say they did. Once they were done talking about their green machines. I followed-up by asking them why they felt the way they did about these two products. Overwhelmingly, they threw out words such as "quality", "brand recognition", "reliability" and "as American as Chevrolet and apple pie." I asked them if they knew how these companies were able to produce these products that they have purchased and enjoy. They struggled to understand where I was leading them with this vague question. I drove my point home by stating that the companies that produce these two products use Continuous Improvement as a means to drive enjoyable, high-quality, reliable, and safe products.

Once I had their attention, I explained to them how Lean Thinking is not something that we only do at work but rather something we also do in our own lives outside work. I provided an example. I explained that after arriving home from work the previous day, I pulled into my driveway to find the outside garage lights on when it was still daylight out. I walked into the basement to find the television on with no one watching it and drinks and food open sitting on the coffee table. I asked them why they thought I was so irritated having to turn several lights off, turn the television off, and put the food and drinks back into the refrigerator. After acknowledging that this happens to them as well, they collectively stated that the cost of electricity and food are not cheap. I explained to them that is exactly my point and this is the mindset we need to have within the company. I went on to state that waste within a company costs money and these costs cannot be passed along to our customers. They reduce company profits which affects us all.

Now that I linked Continuous Improvement to great products and waste reduction / elimination, I provided a few examples of how most humans think with a Lean mindset naturally. I asked them how many of them have ever cleaned out their garages, attics, or basements. Naturally, all of them had at some point in their lives. I walked to the dry-erase board as I was asking them to explain the process they used. As each of them began to call out their steps, I captured them on the board. Once they had completed calling out their steps, I grouped all of their steps into five categories: Sort, Set-in-Order, Shine, Standardize, Sustain, and Safety commonly known within the company as 6S. I went onto ask them how many of them had ever pulled one of their vehicles into their garage to change brakes or had worked on a project around the house like building a deck. As with the 6S example, they all contributed by explaining their process steps. I provided my own example of changing the brakes on my wife's vehicle the weekend prior. After I walked through my process steps, which correlated to theirs, I explained that we refer to this process as Plan-Do-Check-Act (PDCA) within the company.

The 4-hour session was very impactful and was "sticky" for them as I proved the fact, through real-world examples, that Continuous Improvement was more than what the company asks them to do while on the clock. To date, this was one of my favorite training session experiences and I am grateful for all the participants who attended over those two days.

Continuous Improvement also improves safety within processes. I provided 6S training to two export shipping teams while I was working within the locomotive & mining industry. I started the training by asking them to call out their thoughts during the session as I wanted to address them; I wanted the training to be informal and for them to let me know what they were thinking. On several occasions, I addressed their comments which highlighted the fact, for me, that they did not fully understand 6S fundamentals and management was not focused on it. During the 2nd session, there was a gentleman sitting in the back of the room talking to his coworker sitting next to him. He continued to talk as I was speaking. I was curious as to what he was irritated about so I asked him to share his thoughts. He was reluctant at first but eventually stated that he feels that 6S is a joke and that he completes the 6S audit and fills out the end-of-shift 6S checklist but the next shift doesn't do it. He also felt that he does not have time to "tidy" up while he is working doing his assigned job. I thanked him for sharing his thoughts as these are the situations / realities that I wanted to hear and address. I explained to him that my primary focus with 6S is that I want all employees to go back home to their families in the same condition that they left them. I stated that 6S isn't about management being picky about a coffee cup on a tool box but rather about creating a safe work environment where the risk of someone getting hurt has been reduced if not eliminated. I provided a few examples of severe injuries that I personally witnessed as a result of poor 6S practices. In all of these examples, not only were the individuals affected but so were their families. After the session was over, he approached me and asked me for my business card as he stated that all departments within the company needed to receive the training. It became "sticky" with him and the remainder of the group.

The totality of my career has shown me time and time again respect for people, development of people, the power of Continuous Improvement and the value of transformation.

SECTION I | HUMAN EXCELLENCE

Chapter I | PEOPLE CONNECT THE BUSINESS SYSTEM

Lean across a company's value stream is focused on the empowerment of human excellence. I was part of a team who termed the phrase, "People Connect the System", while working within the steel industry in 2008. Years later, I began to refer to it as "People Connect the Business System" as a means of linking people more to the business. As a member of a business system team who were responsible for implementing the company's new business system, it became obvious that the business system only worked when the employees understood it, how it affected their jobs, and how what they did everyday advanced it. What you observe in people is a leading indicator of the future of the operation and therefore the business. When problems do occur, it is important to understand whether the problem is the process or a person.

When people don't understand that their daily actions impact business outcomes, they feel lost and unimportant. This feeling leads to not wanting to contribute or be present.. I heard a saying while in the military… "There is no worse place than the place you are now and no better place than the place where you came from." This saying is based on both bad and good experiences. Companies should not create an environment where the employees don't feel like their contributions are appreciated but rather create an environment where positivity and motivation are wrapped around opportunities for improvement. To add some additional context, let me provide an example of a conversation I had with a newer Procurement employee who was just hired; the company hired her once her internship was completed. She stopped me in the hallway as I was heading to lunch to ask me about some Operational Excellence Training that was being offered the following week. During our conversation, I noticed that she seemed depressed and down. I asked her if everything was alright. She was reluctant to answer, then finally did. She stated that she felt that if she never showed up for work again, no one would even notice. I asked her if she had time to have a working lunch session in one of the breakout rooms. She reluctantly agreed, probably most likely out of curiosity. I used the dry-erase board and started by asking her what were her daily responsibilities. I wrote these tasks down towards the bottom of the board and worked upward showing how these tasks move the needle at the function level which moves the needle at the division level which then moves the needle at the enterprise level. Any work that wasn't moving a needle or foundational for another action was non-value added work. I encouraged her to have a discussion with her manager about increasing the value generated from such tasks. The outcome from this working lunch session was that this person felt like a Navy SEAL walking out of the room on a mission to improve the enterprise. Whether it is one person at a time or a group of people, as leaders, you must ensure that every employee within the company understands the role they play at driving value and how what they do moves the needles.

For organizational transformation to work, the company must work on the culture first. The days of asking employees to leave their brains at the time clocks have long passed. Instead, leaders need to capitalize on the intelligence and experience that the people doing the work can inject into the business. This intelligence and experience transitions into knowledge sharing. Leaders need to understand how to do the jobs that those they supervise are assigned to do. This provides value in multiple ways. There must be an intense focus on the continuous pursuit of human excellence; understanding what makes people more effective and predictive and providing them what they need to be successful. Remember, leaders create the environment that employees succeed or fail within.

W. Edward Deming stated "A bad system will defeat the best people(20)."

Chapter II | CHANGE COMMITMENT & ACCELERATION

The culture that organizational leaders create can make or break a company. The phrase "Culture eats strategy for breakfast" is an accurate statement. Culture should be about the company's shared beliefs of how to successfully run the business.

All that you have learned in this book is useless unless the people within the organization understand it, are included in it and are committed to driving it. This is such a simple component that most companies are not successful leveraging. Sure, most companies talk about it but do they really live it? Is it in their DNA?

The objective is to drive commitment and not compliance. The people who are compliant are not being compliant because they are bad employees or want to come to work and do the least amount of work possible. The primary reason is that at some point in time, change was introduced within their process and the creator of the change did not include them in the change process.

Everyone resists change at some point of time within their careers. People, like companies, lack the patience to embrace change. Listed below are some questions people have when faced with change:
- What's in it for me (WIFM)
- How will my job change
- Will I be moved somewhere else within the business
- Will I be required to use technology... I have never had to use it before
- Will I lose money
- Will I no longer be working with the team members I have been working with for over a decade
- What's the risk if it doesn't work out

All these questions are understandable. I have asked some of these same questions early in my career when faced with change. The factor generating this panic and concern for people is that their input was not solicited as the business case was being created for the change. Employees should not be blindsided by change. In these instances, change is forced upon employees with limited and even in some circumstances no explanation other than "this is what we are doing." When people's input is part of the change and they understand the business case behind the change, the majority of them will be committed and not just compliant. To validate this point, most people are not affected at all with change in their personal lives; outside the four walls of the company. In many cases, they will actually pay for change. Let me provide an example that I provided during Change Commitment & Acceleration training I provided at the NASA Stennis Space Center:

I kicked off the training by asking the room of highly intellectual people how many of them had the newest cell phone. The majority of them raised their hands. I called on someone and asked them if they upgraded their cell phone because the one they had no longer operated. They responded by stating no and that their old cell phone still worked. I then asked them why they chose to pay money to upgrade the cell phone when the one they had still worked for them. The response was because they wanted to stay up-to-date with the technology that their friends and colleagues were using and wanted the additional functionality even though they may not use it. I asked the same question of the group with respect to flat screen televisions and received a similar response. The message that I was driving was that change does not bother us in our personal lives but when asked to change within the companies we work for, we don't like it. If we begin to unpack this thinking, we will discover that the changes we make in our personal lives are mostly driven by us whereas change within the companies we work for are driven by management or senior leadership. Most of the time, people were not asked for their input.

So what is **Change Commitment & Acceleration**, you may be wondering. It is simply a model that can be used to drive commitment of a change and accelerate that change. It becomes a mindset for creating and welcoming change. That change becomes opportunity for companies. The model is comprised of a set of tools that transitions compliant behavior to committed behavior. This is a valuable model for anyone who is seeking to introduce change within the organization. **(Figure 2-1)**

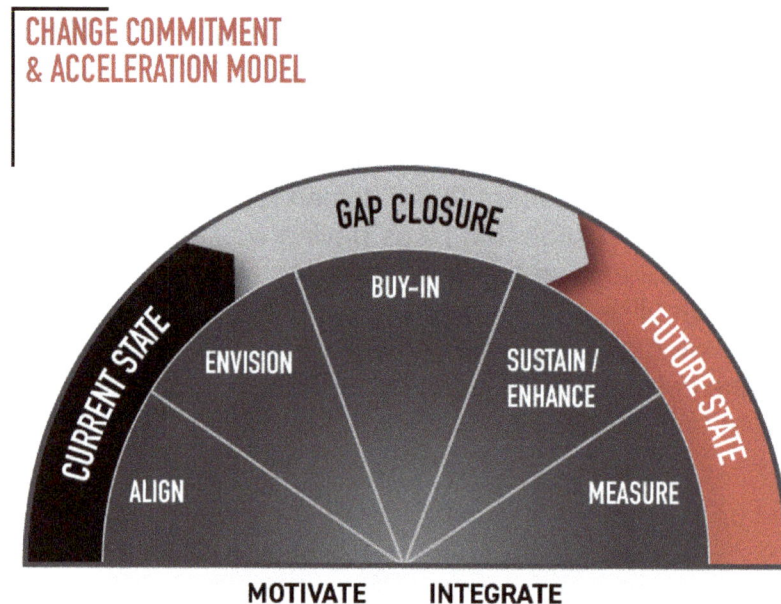

CHANGE COMMITMENT & ACCELERATION MODEL

I learned, while working at GE, that while 100% of successful changes have a good technical solution, so do 98% of unsuccessful changes. The difference is commitment.

How this model works is as follows:
MOTIVATE CHANGE:
  • Generate a level of excitement about the "what is possible". Not everyone is capable of doing this so select the right person.

Success Achieved When:
  • Changes are selected strategically and drive the organizations' strategic plan.
  • Teams are formed with a clear definition of roles and expectations.

24

Where Failure Occurs:
- Upfront motivation is inefficient and thus does not drive the level of interest and excitement to warrant support.
- The change does not drive the organization's strategic plan neither directly nor indirectly.
- Lack of direction for the formed teams.

ALIGN THE NEED FOR CHANGE:
- Create the business case (always begins with the "why") for change by doing the due diligence to ensure that it is well thought out and constructed with a high level of intelligence behind it. Simulate what possible questions and / or concerns the people who would be affected by the change would have.

Success Achieved When:
- The business case for change is effectively delivered and input is requested from those who will be affected by the change; it provides a sense of ownership… "skin in the game".
- The business case is adjusted to reflect any input, where applicable, and everyone understands it and is supportive of it.

Where Failure Occurs:
- Assuming the need (the "why") for the change is clearly articulated.
- The business case is not well constructed.
- Not requesting input.
- Not understanding that not everyone will agree with your definition of the problem for which you are proposing a change to correct.

ENVISION THE CHANGE:
- Create the vision statement that communicates the targeted outcome of the business case for change. This is the opportunity to generate additional motivation and excitement about the change.

Success Achieved When:
- The business case for change is energizing and motivating.
- Understood by all.
- Measurable.

Where Failure Occurs:
- The person delivering the vision is not motivating.
- When there is multiple versions of the vision.
- Too high level.
- Too complicated.
- The vision does not address the specific needs of those affected by it.

BUY-IN OF THE CHANGE:
- Gain commitment of those affected by the change. If resistance exists, understand where the resistance is and what it is and work to reduce and / or eliminate it. I recommend reading *Ten Reasons People Resist Change* by Rosabeth Moss Kanter, in which she explained each of the following reasons listed here(21).

  1. Loss of control
  2. Excess uncertainty
  3. Surprise, surprise!
  4. Everything seems different
  5. Loss of face
  6. Concerns about competence
  7. More work
  8. Ripple effects
  9. Past resentments
  10. Sometimes the threat is real

Success Achieved When:
- Identifying the people who the change would affect.
- Understand the different types of resistance.
- Understanding how to achieve a "win / win" outcome.
- Implementing strategies to build and sustain commitment.

Where Failure Occurs:
- Not understanding where the resistance is.
- Poor communication.
- Assuming a technical or skill-related solution is sufficient.
- Not fully understanding the political element.
- Not acknowledging and managing the social and human aspects of the change.

SUSTAIN / ENHANCE THE CHANGE:
- Celebrate success. Monitor and reinforce the vision to ensure new behaviors do not transition back to old behaviors ("mission creep"). Work with those who had input into the business case for change to seek ways to enhance, where applicable.

Success Achieved When:
- Keeping the change exciting.
- Communicating the early wins to create momentum.
- Not becoming complacent with the change and enhancing when something within the internal or external environment changes.

Where Failure Occurs:
- Permitting the change to lose priority to other initiatives for time, resources, or focus.
- Losing change momentum.
- Organization falling into a state of "mission creep".

MEASURE THE CHANGE:
- If people cannot measure the change, then they cannot control it; they don't know if they are winning or losing. An accurate measure of the change provides focus, alignment, direction, and momentum.

Success Achieved When:
- Key Performance Indicators (KPIs) are used to drive accountability.
- KPIs are used to determine if milestones have been achieved and celebrated.

Where Failure Occurs:
- KPIs are not used or if used, not monitored and communicated.
- Accountability is not maintained.
- Data are not clean.
- Outcomes not communicated to the stakeholders.
- Not using the correct KPIs.

INTEGRATE THE CHANGE:
- Adapt systems and organizational structures to reinforce and support the change and enhancement improvements.

Success Achieved When:
- The change becomes part of the organization's daily management; the how they run the business.

Where Failure Occurs:
- A change in leadership sidelines the change.
- Systems nor adapted.

There are tools per component of the model that are used to engage commitment, reinforce behaviors that support change within a business environment, and accelerate that change. **(Figure 2-2)**

CHANGE COMMITMENT & ACCELERATION

HOW

Tool	Stage	Description
VIEWFINDER	MOTIVATE	Brainstorm the impacts, resources, and results that should or should not be included in the scope of the change.
16-WORD CANVAS	MOTIVATE	Define the change by writing a sufficient definition.
R&R MATRIX	ALIGN	Identify the potential rewards as well as the inherent risks in making the change.
"PROVE IT" MATRIX	ALIGN	Identify objective sources that support the need to change; some you may have in hand and some you need to find.
REARVIEW MIRROR	ENVISION	Describe a potential future state in terms of environment, mindset, results, and customers.
MORE OF / LESS OF	ENVISION	Identify specific behaviors that are supportive (More of) and behaviors that are not supportive (Less of).
PITCH	ENVISION	Write a pitch to communicate the change and its needs.
STAKEHOLDER ANALYSIS	BUY-IN	Identify stakeholders, target behaviors, and the influence needed to support the change.

I have used the Change Commitment & Acceleration Model and associated tools on many occasions within my career. I have used the Stakeholder Analysis Tool to better understand what type of resistance I was facing: was it political, technical (most common), or cultural (most challenging). Understanding the source of resistance helped me to mitigate the risk through the approach I took to address it. Within the first 90 days of being in a new role, I leverage this tool to better understand and identify who my supporters are and who those who are strongly against me. **(Figure 2-3 - 2-3d)**

STAKEHOLDER ANALYSIS TOOL
Who Are Your Stakeholders?

								TEMPLATE

STAKEHOLDER NAME	AREA IMPACTED	SUMMARIZE CHANGES IN EXISTING WORK PROCESSES WILL OCCUR.	STRONG AGAINST	MODERATELY AGAINST	NEAUTRAL	MODERATELY SUPPORTIVE	STRONGLY SUPPORTIVE	SUMMARIZE REASONS FOR RATING
(WHO)	(WHERE)	(WHAT)	(RATING 1, 2)	(RATING 3, 4)	(RATING 5, 6)	(RATING 7, 8)	(RATING 9, 10)	(WHY)

Legend: C = Current D = Desired

			ILLUSTRATIVE EXAMPLE

STAKEHOLDER NAME	STRONGLY AGAINST		AGAINST		NUETRAL		SUPPORTS		STRONGLY SUPPORTS		REASON FOR RATING
RATING	1	2	3	4	5	6	7	8	9	10	
Person 1	C							D			Gives the perception that they are supportive but secretly undermines the initiative.
Person 2			C				D				Didn't attend 4 meetings.
Person 3						C				D	Attends meetings but doesn't often offer any input or ideas.

STAKEHOLDER ANALYSIS TOOL
What is Their Resistance?

			TEMPLATE

STAKEHOLDER NAME	TECHNICAL	POLITICAL	CULTURAL
(LAST, NAME, FIRST)	(FEAR THEY LACK SKILL, FEAR OF UNKNOWN, HABIT, INVESTMENT IN OLD WAY, PERSONAL FEAR OR DOUBT)	(LOSS OF POWER, CONTROL, DECISION MAKING ABILITY, STATUS, SELF-PERCEPTION, TURF OR RELATIONSHIP STRUGGLES, WHO GETS TO TALK TO WHOM)	(DIFFERENT THAN HOW WE DO THINGS HERE, OLD MINDSETS, GOOD OLD DAYS, BLINDERS, FEAR OF LETTING GO)

			ILLUSTRATIVE EXAMPLE

STAKEHOLDER NAME	TECHNICAL	POLITICAL	CULTURAL
Person 1	I am very comfortable and efficient with the tools.	I should get credit for what I've done. She has done little.	
Person 2		I manage over 250 people; I don't want to lose influence over my team.	I've been here longer and I am more well rounded than everybody else.

STAKEHOLDER ANALYSIS TOOL
What are The Targeted Behaviors?

							TEMPLATE
STAKEHOLDER NAME	LEADS, PARTICIPATES, SUPPORTS, INFORMED OF, IMPACTED BY THE CHANGE?	EXPECTED IMPACT OF CHANGE TO THEIR AREA	LEVEL OF INFLUENCE	DEFINED BEHAVIOR	BEHAVIOR	ISSUES / CONCERNS	IDENTIFY "WIN / WIN"
(LAST, NAME, FIRST) (WHO)		(H.M.L.)	(H.M.L.)		(MORE OR LESS OF ?)	(PERSONAL THREATS)	(PERSONAL OPPORTUNITIES)

				ILLUSTRATIVE EXAMPLE
STAKEHOLDER NAME	DEFINED BEHAVIOR	BEHAVIOR (MORE OF / LESS OF ?)	ISSUES / CONCERNS (PERSONAL THREATS)	IDENTIFY "WIN / WIN" (PERSONAL OPPORTUNITIES)
Person 1	Disparaging a critical path tool.	Less of	People see me as less intelligent because I am less efficient with another tool.	Realized efficiency with new tool.
Person 2	Advising new hires.	More of	Manage less people.	Be a future-focused / thought leader.

STAKEHOLDER ANALYSIS TOOL
What are The Next Steps?

				TEMPLATE
STAKEHOLDER NAME	INFLUENCE STRATEGY TASK	INFLUENCE STRATEGY ASSIGNED TO	INFLUENCE STRATEGY BY WHEN	INFLUENCE STRATEGY MEASURE
(LAST, NAME, FIRST)				

				ILLUSTRATIVE EXAMPLE
STAKEHOLDERS	TASK	ASSIGNED TO	BY WHEN	MEASURE
Person 1	Meeting with Tom to provide training	Jerry	04/17/23 1:30pm	Tom can teach it to others.
Person 2	Scheduling mentorship meetings	Nick	07/11/23 10:30am	

Change Commitment & Acceleration is not to be confused with change management. Change management is an approach to deal with change that has already occurred. Change management is focused on defining and implementing procedures and / or technologies to adapt to, control, and affect change within the organization. It is not focused on gaining commitment or acceptance of change.

Simply put, commitment from people is what will ensure that everyone in the organization is pushing in the same direction. At the end of the day, this is what any organization desires.

SECTION II | BUSINESS IS WAR

Chapter III | COMPETITIVE ADVANTAGE

Competitive advantage is a condition; a dynamic state that positions a company to produce a good or service of equal value at a higher perceived value; the customer value proposition. This is accomplished through high-quality goods and services that provide leverage over their competitors. It comes from approaching and doing things differently. This condition / state enables the company to generate more sales thus driving improved margins compared to its market competitors. Competitive advantage is accomplished through a defined strategy. Strategy provides choices and options. Companies must constantly ask themselves what is the value of not doing something? What is the value of continuing what we are currently doing? Over time, competitive advantage begins to dissipate as a result of reliance on and complacency with a single approach.

It is essential for companies to have a battlefield mindset. Companies must keep focused on their customers but always aware of their competition. It's about outsmarting the competition. I recall hearing, while working for Sony, that Akio Morita, CEO of Sony, stated that *"Business is war."*

In their book, *Playing to Win(3),* A.G. Lafley and RL Martin stated:

> *"I have found that most leaders do not like to make choices. They'd rather keep their options open. Choices force their hands, pin them down, and generate an uncomfortable degree of personal risk. I've also found that few leaders can truly define winning. They generally speak of short-term financial measures or a simple share of a narrowly defined market. In effect, by thinking about options instead of choices and failing to define winning robustly, these leaders choose to play but not to win. They wind up settling for average industry results at best."*

Recall what I stated in the Introduction Chapter…Simplify and Control enables the organization to disrupt the game whereas Proactive and Predictive enables the organization to dictate the game. Disrupting and dictating is about winning. Utilizing the 4 Phase Approach Model is about making that choice to win.

The strategic logic of competitive advantage is comprised of:
1. Purpose- what is the company's winning aspiration
2. Scope- where will the company play
3. Internal organization- what activities, investments, and capabilities will support the company's strategy

4. External organization- how will the company strategically position itself to achieve its objective

To create competitive advantage across any industry, organizational size, and product or service, Porter's Generic Strategies can be used. These Generic Strategies were first introduced by Michael Porter in 1985 through release of his book titled: *Competitive Advantage: Creating and Sustaining Superior Performance(4).* **(Figure 3-1)**

PORTER'S GENERIC STRATEGIES

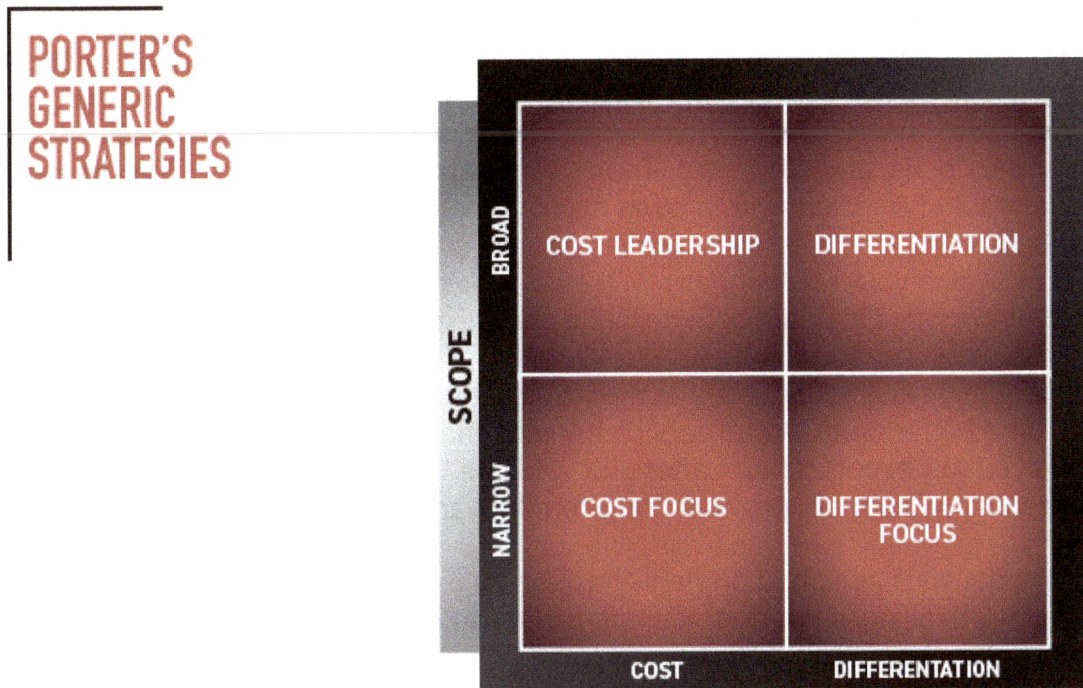

Porter categorized the Generic Strategies (ways of winning) as:
1. Cost Leadership
2. Differentiation
3. Cost Focus
4. Differentiation Focus

Cost leadership is a focus on cost… a company wins through cost. This is accomplished through cost reduction which results in lower prices at reasonable perceived value which increases profit and market share. Standardization becomes a core driver for cost leadership. A cost-based competitive advantage normally provides short-term advantages.

The primary driver for organizational cost reduction is a focus on Operational Excellence. Continuous Improvement is the vehicle for Operational Excellence and used by most companies. You may be thinking if this is used by most companies, then where is the competitive advantage? If all companies have access to Continuous Improvement and can easily copy and paste someone else's cost reduction playbook, then why is Operational Excellence not always achieved? The key is the information detailed in this book. Let me explain… from my nearly three decades of experience, most companies talk about Continuous Improvement in shareholders reports, to the Board, and to the Wall Street analysts. They have amazing posters on the walls and spend marketing and public affairs funds on producing great business, operating, and production system content but they simply do not do what they say they will do for many reasons that I have detailed in this book. There is a lack of organizational discipline, focus, and alignment. The 4 Phase Approach Model provides an easy to understand structure with proven results. The advantage of this book for the reader is it provides where companies can fail which is what my experience is now providing you. Knowing this will remove many obstacles along a company's journey towards Operational Excellence and thus competitive advantage.

Differentiation is what sets one company apart from their competitors. It means providing better product and service benefits that are different from and perceived to be more sought after and valuable than their competitors. Differentiation normally provides long-term advantages.

Differentiation is launched primarily through innovation (breakthrough / disruptive thinking… the "do different"). The five components of innovation are:
1. Empathize- Identify the needs
2. Frame- Identify the opportunity
3. Ideate- Identify the possibilities (idea generation)
4. Create- Build the concepts and stories
5. Test- Ensure it solves the needs

Cost Focus and Differentiation Focus are simply focus on either cost leadership or product and /or service differentiation; general focus is not as effective and adequate enough. These strategies are used more so to target niche markets and do so better than anyone else. Often times, the company is focused on a niche that is not currently being served.
Fulfilling the needs of niche markets enables the company to gain brand loyalty where customers are willing to pay more.

Note: some companies fail to engage one of the 4 Generic Strategies. When this occurs, the company is said to be "stuck in the middle". An indication of this state is when a company does not offer product and service features that are unique to generate customer interest and its prices are set too high to effectively compete based on price alone. The probability of failure is high on this position. Porter stated *that the profitability of firms depends not only on the typical rates of return in an industry. It depends more importantly on the firm's position and competitive advantage in that industry(4).*

In addition to Porter's Generic Strategies, he is the architect of Porter's Six Forces Model. This model provides a framework that identifies and analyzes the primary sources of competition within industry. It characterizes the competitive environment, not a specific company's capability. **(Figure 3-2)**

PORTER'S SIX FORCES MODEL

THREAT OF NEW ENTRY

POWER OF SUPPLIERS

COMPETITIVE RIVALRY

POWER OF COMPLIMENTS

THREAT OF SUBSTITUTION

POWER OF BUYERS

Company competitive advantage comes in many forms including but not limited to:
- Products
- Services
- Marketing Channels / Positioning
- Distribution Network
- Branding / Brand Recognition / Customer Loyalty
- Quality of Offerings
- Customer Service

- Strategic Pricing
- Company Reputation
- Innovation and Access to New Technologies
- Intellectual Property
- Operating Model
- Investment and Development of Employees

The most overlooked forms of competitive advantage are:
- Breakthrough Management (Execution of developed strategy)- developing strategy and its associated desired financial / non-financial outcomes are somewhat difficult at times but easier than executing said strategy. Using Hoshin Kanri (Japanese for policy management), also known as strategy deployment, or as I have referred to it, Strategy to Execution (STE), is an effective structure / framework to keep the organization focused, aligned, and prioritized on successful strategy execution. Structure always follows strategy. STE is that structure.

- Change Commitment & Acceleration (Committed culture to drive the execution)- having a committed culture as compared to a compliant culture is critical for any organization.

- Daily Management & Lean Management (Continuous Improvement)- focus on customer value, waste identification and reduction / elimination, risk, creating a problem predictor culture, and lessons learned. Jack Welch (Former CEO of General Electric) stated that *an organization's ability to learn and rapidly translate that learning into action is the ultimate competitive advantage(5).*

The Model uses Human Excellence, Change Commitment & Acceleration, Breakthrough Management, Daily Management, and Lean Management to drive and support the Model Phases. Execution of these Model Phases contribute to Porter's Four General Strategies specifically Cost Focus and Cost Leadership.

Competition drives the need for competitive advantage. Competition is about creating change. And change drives opportunity. A company's objective should always be to remain relevant. In business, as it is in life, relevance is everything. A company's objective is to remain relevant. Companies need to always be transforming.

SECTION III | THE GENESIS OF THE MODEL

Chapter IV | THE EXPERIENCE FACTOR

One thing that my experience has provided me is insight into many of the common problems that affect most companies irrelevant of the industry. I have worked within public, private, and non-profit companies to include government. Industries that I have gained experience from within the manufacturing sector are: Electronics, Bottling, Metals, Rail & Mining, Energy, Automotive, Specialty Automotive. Industries that I have gained experience from within the service sector are: Financial Services, Transportation & Logistics, and Aerospace & Defense. Within the manufacturing sector, I have worked for companies who manufacture and process high volume / low cost product, low volume / high cost product, standard platform product, and customized product within assembly line and job shop environments enabled by both union and non-union teams. My insight developed while I was working within the metals industry. At that time, I created an Ishakawa Diagram, also referred to as a Cause & Effect Diagram, but most commonly known as a Fishbone Diagram. I listed all the common organizational problems that applied across all of the companies I had worked for. **(Figure 4-1)**

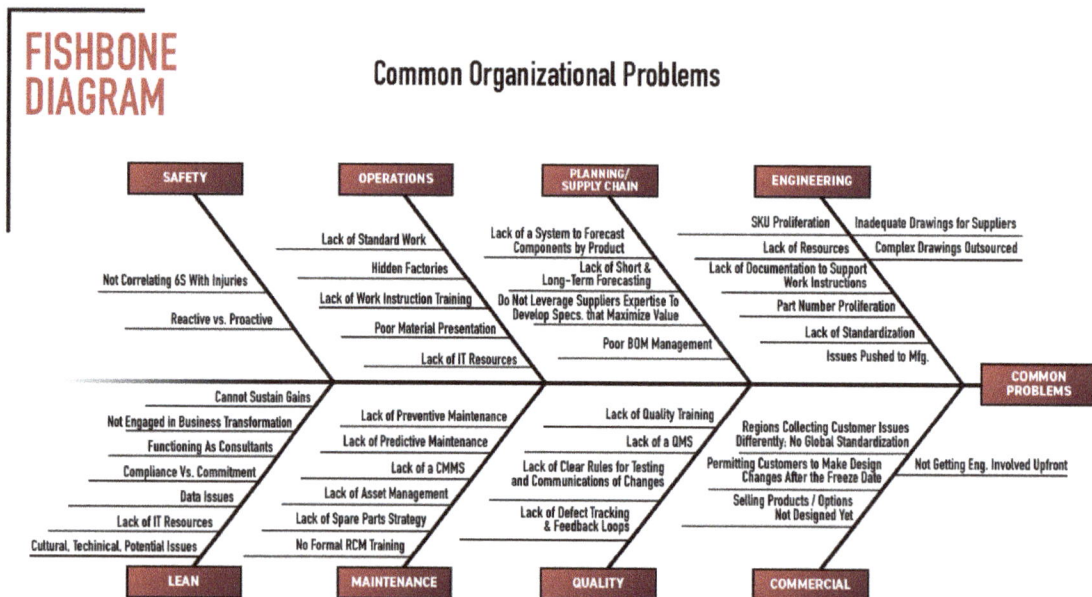

FISHBONE DIAGRAM

Common Organizational Problems

I continued to keep it updated as I gained additional experience within other companies and industries. This experience eventually led to the creation and validation of the 4 Phase Approach Model. It is a common sense approach that

x

applies to any business' value stream and more importantly to the common problems that affect most companies irrelevant of industry.

Experience provides lessons learned and lessons learned lead to innovative thinking, such as creating hybrid Lean tools for a specific purpose. An example of this is when I combined a Cause and Effect Diagram with a Check Sheet and created the Cause and Effect Check Sheet. **(Figure 4-2)**

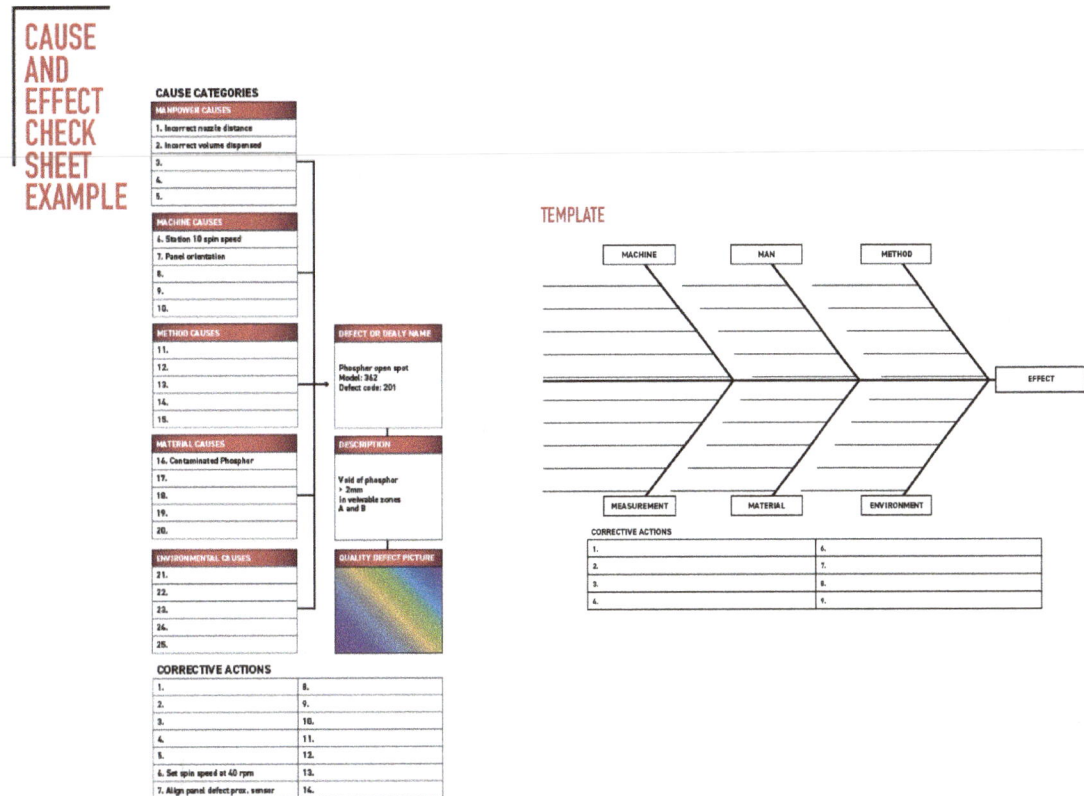

This hybrid tool was made public through the American Society for Quality which was included in the March 2014 edition titled: *At Your Fingertips*. This was my second published article. I was working within a processing plant where our top two quality defects resulted in a ~$600,000 cost per year. The frequent occurrences of the defects swung the Quality Factor (QF) needle in the wrong direction and drove down the Overall Equipment Effectiveness (OEE) metric. The OEE metric provides the percentage of an asset's capability that is being used after the cumulative effect of the three major losses (delays, speed, and quality). In addition, in many occasions, the antecedent process feeding into the process in question had to be idled or even shutdown depending on if any of the machines were damaged as a result of the defect event. I worked with the process supervisors to analyze the data and present it in a Pareto Diagram. From the diagram, we decided to focus on the top five products (based on frequency and cost impact). From this point, I created a Fishbone Diagram for each product where I had the product name as the head of the diagram and listed all the historic causes from each of the two defects (each of the five products had two separate diagrams, one for each defect) off of the spine of the diagram. I now looked at what the solution was for each of the causes that were known. I created a box at the bottom of the diagram where I listed and linked all solutions to each of the causes. These Cause and Effect Check Sheets were then laminated and placed in a pre-product run binder. The supervisors would use the Check Sheets, based on the product that was scheduled next to be processed, as a check sheet to confirm, for example, equipment settings. As solutions were identified, the Check Sheets were updated; they became living documents. Within a three month period, the OEE metric improved by 11% for the one defect and by 8% for the second defect. The overall cost savings was $430,000. **(Figure 4-3)**

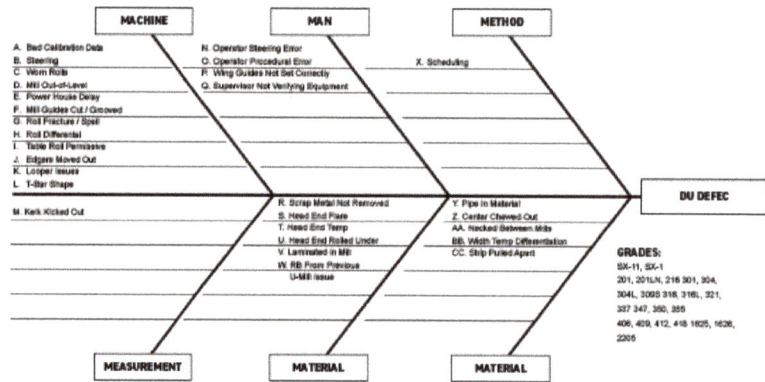

CAUSE AND EFFECT CHECK SHEET EXAMPLE

I am very fortunate to have had several sensei (mentors) over the years. Each sensei has roots directly or indirectly back to Toyota with the Toyota Way and Toyota Production System. I was mentored for 11 years while working for Sony, mentored for two and a half years by Shingijutsu, USA sensei while working for GE, and mentored for 4 years by a sensei at Lean Sensei International while working at Global Container Terminals. Lineage back to the source is always an important component of learning. Even though Toyota did not discover Lean, they did leverage it, enhanced it, and embedded it within their culture and DNA that then became a force multiplier for success for them within the automotive industry. As an example, if you wanted to learn Brazilian Jiu Jitsu (martial art that actually originated in Japan), ideally you would want to learn from a Brazilian teacher specifically someone from the Gracie Family. The Gracie Family learned Jiu Jitsu from the Japanese, then enhanced it and took the world by storm. I learned Jiu Jitsu from a Brazilian teacher who learned directly from a Gracie teacher. This authentic lineage has proven to be very valuable for me with respect to my Jiu Jitsu journey.

Over the years as I have interviewed with companies for Operational Excellence leader roles, I have been asked if I would capture my strategy and approach on one to two slides. This would provide them some insight into whether I would be too much of a high-level / strategic leader or a "roll up my sleeves" more tactical leader. In addition, they could ascertain from the information whether I would function as a strategic business partner to those who I would be supporting rather than a consultant who would simply tell them what's wrong. My years of experience working within Operations leveraging Operational Excellence and working within Operational Excellence supporting Operations has provided me an effective strategy and approach that I have successfully used across multiple companies. The strategy is listed below and the approach used to drive that strategy is The 4 Phase Approach. **(Figure 4-4)**

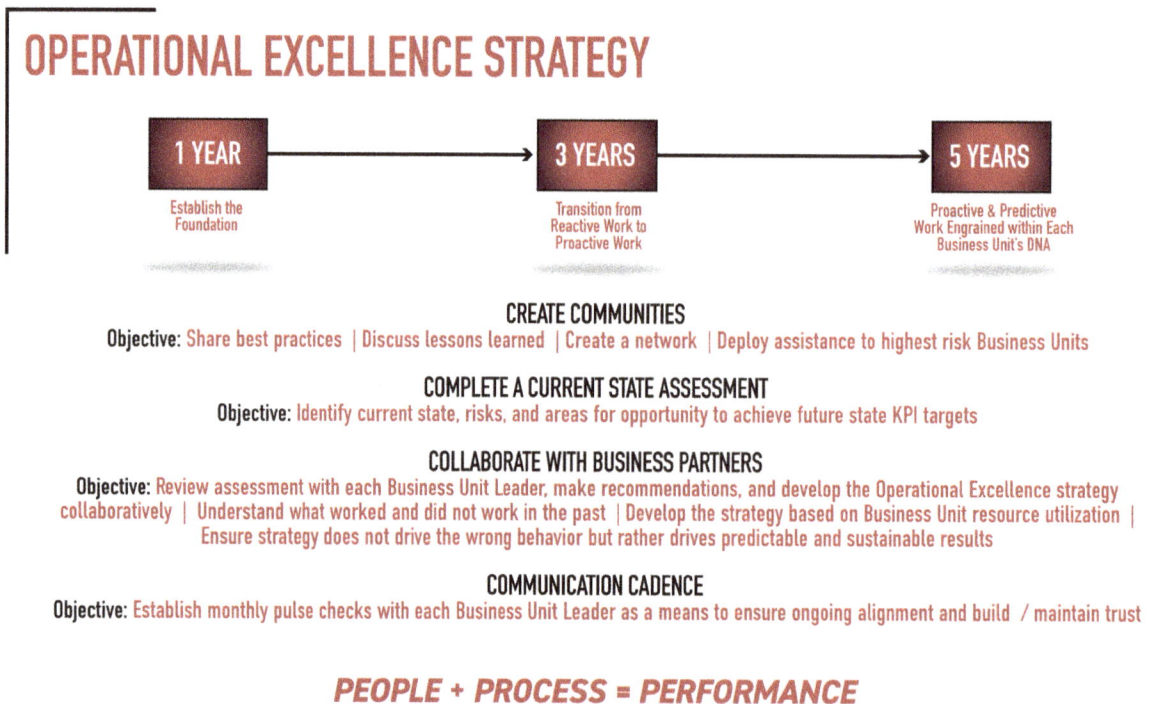

OPERATIONAL EXCELLENCE STRATEGY

1 YEAR	3 YEARS	5 YEARS
Establish the Foundation	Transition from Reactive Work to Proactive Work	Proactive & Predictive Work Engrained within Each Business Unit's DNA

CREATE COMMUNITIES
Objective: Share best practices | Discuss lessons learned | Create a network | Deploy assistance to highest risk Business Units

COMPLETE A CURRENT STATE ASSESSMENT
Objective: Identify current state, risks, and areas for opportunity to achieve future state KPI targets

COLLABORATE WITH BUSINESS PARTNERS
Objective: Review assessment with each Business Unit Leader, make recommendations, and develop the Operational Excellence strategy collaboratively | Understand what worked and did not work in the past | Develop the strategy based on Business Unit resource utilization | Ensure strategy does not drive the wrong behavior but rather drives predictable and sustainable results

COMMUNICATION CADENCE
Objective: Establish monthly pulse checks with each Business Unit Leader as a means to ensure ongoing alignment and build / maintain trust

PEOPLE + PROCESS = PERFORMANCE

My focus has always been on how to create repeatable / sustainable processes and results through team-based / data driven continuous improvement. This learning has been a result of reflecting back on the failures and successes with Lean. This reflection has provided me knowledge but more importantly wisdom. The transformation journey is about the people and process. For this reason, I begin with Human Excellence which underpins the 4 Phase Approach Model.

Chapter V | SIMPLIFY

In 1492, Leonardo DaVinci said *"Simplicity is the ultimate sophistication(6)."* Einstein stated *"If you cannot explain it simply, you don't understand it well enough(7)."* Interestingly enough, organizational knowledge leads a company from simple to complex; organizational wisdom leads the company from complex to simple as I stated in a previous chapter. Lean is a catalyst and driver for simplification but if not guided and implemented correctly can create unwanted complexity. Simplify defines the way we make decisions, develop and commonize / standardize processes, work together, and work with our partners and customers. It is simply the way we should run the business especially during periods of transformation. It sets the foundation; it becomes linear from which standards can be applied. It is focused on rationalizing and reducing / eliminating non-value added activities. Simplifying is an approach that forces the organization to look at each area of the value stream and understand each step of each process and where wastes and opportunities are apparent. It must begin with observation and identification. Hidden factories are a source of waste. When there is an absence of standard work, resources (materials, tools, etc.), technical support, and leadership in addition to excess inventory, employees will create the end arounds that become the hidden factories. These end arounds give off hidden signals and are not factored into organizations risk models. Always remember that wastes do not create value for the customer, they create organizational inefficiencies which works against flow. It is important to understand what is value first. Whatever is not value is waste. There are eight types of wastes:
1. **D**efects
2. **O**ver Production or Processing
3. **W**aiting
4. **N**ot utilizing Peoples Skills
5. **T**ransportation
6. **I**nventory
7. **M**otion
8. **E**xtra Processing

There are three contributors for the eight wastes:
1. Not adding value (Japanese word is Muda)
2. Unevenness (Japanese word is Mura)
3. Overburden (Japanese word is Muri)

To simplify does require a change in culture as it will drive people to make decisions where the work is occurring. This required culture change will vary based on the current maturity state of the organization. To simplify will result in the removal of complexities by identifying and removing the friction out of the processes and commonizing / standardizing them resulting in a gain in organizational speed. This is most relevant with factories containing multiple product lines. When new intelligent factories are being designed and created, simplification is built-in. And always remember to include the voice of the employee (VOE), a term I created back in the mid 90's while working on an assembly line. Each time I have worked to design and develop a factory, the team was guided by a structured approach such as the Production Preparation Process (3P). This tool is one of many that are only effective when the VOE is involved.

Complexity slows down organizations. Complexities develop within organizations for several reasons but my experience has shown me that two of the primary reasons are lack of communication and silos. The complexity lies within the handoffs between the silos. Leaders must manage and remove complexity to ensure continuous improvement and forward progression of the organization; to achieve the True North. Organizations must focus on creating simplified processes that work for employees who make the products and / or provide the services.

Lean drives simplification. Lean techniques are proven to improve safety, quality, employee retention, delivery, cost reduction, efficiencies, flow, and even identify new revenue streams. Lean must be a critical part of the organization's strategy but organizations must be willing to step out of the reactive work into Lean Thinking (focus) in order to achieve results. Lean focus is required for organizations within times of economic uncertainty such as the 2008 Stock Market Crash and the 2020 COVID-19 Global Pandemic to best manage cost and mitigate risk.

I held the position of a Business System Leader (Lean Leader) within the metals industry early on in my career during the 2008 Stock Market Crash. I recall the Lean Team being called to the President's office: I packed my office up and placed the boxes in my car prior to heading to the meeting. I thought that due to the reduction of volume the company was experiencing and forecasting, the Lean Function would be the first to be let go. My perception was that teams such as marketing and operations were felt to be essential to generate revenue but Lean functions could be eliminated and not affect the forward progress of the company. To my surprise, the president stated that of all times, he needed us to manage the middle of the income statement as the top line (revenue) was not going to improve and if anything, it would decrease. He stated that we have the opportunity to push cost out (variable) of the business and thus increase the bottom line (profit). I walked out of that meeting feeling like a Navy SEAL heading into battle.

During the 2020 COVID-19 Global Pandemic, I was leading a global Lean Function within the transportation & logistics industry. The focus was on ensuring that the company had a strong Free Cash Flow (FCF) which comes from managing variable costs, limiting investments in Capital Expenses (CapEx), and managing asset availability and reliability through a Reliability Centered Maintenance (RCM) program. With the uncertainty of how long the pandemic would last and the risk associated with it, the investors were relying on the organization to manage the financials appropriately. The Lean Function, working with the organization as strategic business partners, worked diligently to ensure that we were not leaving anything on the table with respect to cost leadership.

Tim Peters, The Zen of Python, stated *"Simple is better than complex. Complex is better than complicated(8)."* Jeff Immelt (then CEO of General Electric) wrote to shareholders in early 2016 stating *"In a complicated world, you need to be simpler and more competitive(9)."*

Caution Points for Simplifying:
Be careful not to over-simplify to a point where you begin to create inefficiencies. A major cellular provider thought they simplified their in-store account updates through use of a different software system but in reality, it took an employee more time to complete transactions resulting in additional time for the customer to wait. A major national bank thought they would simplify their check-in process for other than bank teller transactions by having a QR code attached to the bank greeter podium. The greeter would ask customers to scan the code with their electronic device then answer a few questions in order to be placed in the waiting queue. The problem was that they did not account for people who either did not have or have with them their cellular devices. Needless to say, this process lasted less than 30 days and was eliminated.

These examples illustrate why it is important to simplify, but check to ensure you haven't inadvertently introduced complexity. One of my senseis once told me to think creatively, don't simply throw money at it; focus on removing the monuments from the genba (Japanese for place where work occurs). He defined a monument as anything that could not be moved or corrected within a weekend time frame. Adopting an organizational Lean mindset, prior to implementing Lean, is a must. This can be accomplished through Lean training, leaders who help all employees see how their contributions move the needle, and servant leadership. As discussed in Chapter II Change Commitment & Acceleration, most importantly is when change is required, leaders must ensure that they do their due diligence creating the business case for change and present it to those employees who will be affected by the change and ask for their input. These simple steps transition compliance to commitment as the affected employees now have skin in the game; they are part of the decision making.

SIMPLIFY TOOLS
1. Value Stream Mapping (VSM)
2. 6S
3. Production Preparation Process (3P)
4. Disruptive Cost Out (DCO)
5. Visual Management
6. Work Instructions

1. Value Stream Mapping (VSM):
Everything is a system comprised of individual linked components. If you look hard enough, you will see the links and connection points within a business. Think of a value stream as a system or a connected ecosystem. In order to simplify something, you must be able to see it; you must be able to visualize it. Simplification begins with making visible and understanding the organizational value stream's current state. In absence of this, the organization cannot see the handoffs, rework loops, hidden factories, "white space," and risk and thus opportunities cannot be identified. Simplification is accomplished through Value Stream Mapping (VSM).

This tool is categorized in the Simplify Phase as making a current state process visible where it can't otherwise be seen to this degree, with the goal of creating a more efficient future state, is the first step in process improvement.

A VSM is a graphical representation of the key process steps in an organizational value stream. It is a diagnostic tool; a philosophy on how to approach improvement. Understanding and defining how the customer views value is the first of the Five Lean Principles detailed in the book *Lean Thinking* by James P. Womack and Daniel T. Jones(10). **(Figure 5-1)**

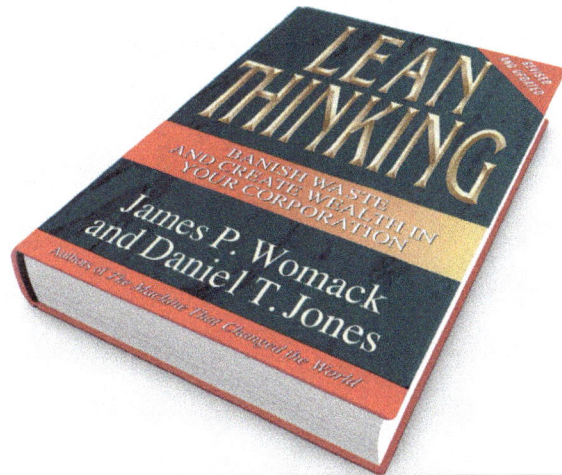

Second-Value. Mapping a Value Stream...the way value is delivered.
VSM helps us understand processes and is a common language for quickly assessing relevant process steps and their connections. It is a structured approach to finding the "noise" and identifying the opportunities in processes focused on holistic improvement(11). **(Figure 5-2 & 5-2b)**

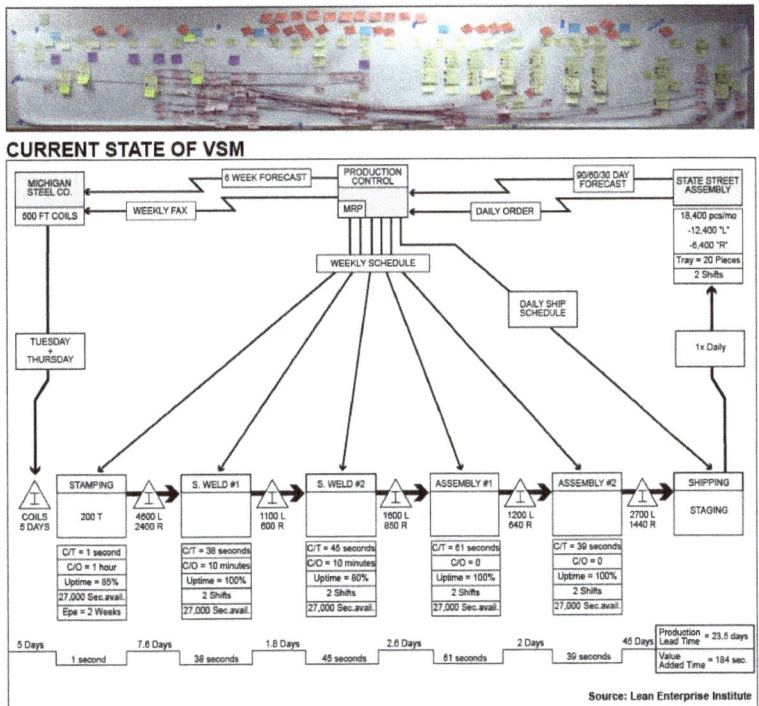

CURRENT STATE OF VSM

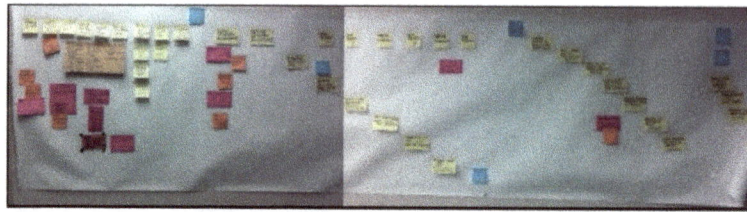

FUTURE STATE OF VSM

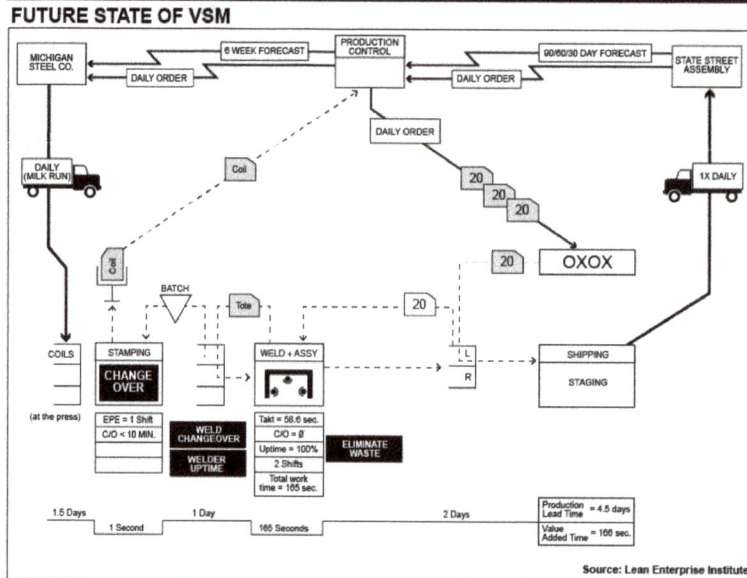

Source: Lean Enterprise Institute

In further detail, a VSM is a graphical representation between material and informational flow of the key process steps, including value added and non-value added actions. VSM is used to deliver a product or service to an internal or external customer. It is compromised of all of the steps, both value-added (VA) and non value-added (NVA), required to take a product or service from raw material to the customer. VA is anything that the customer is willing to pay for, that changes the value of the object or service, and that is processed only once.

The objective of a VSM is to optimize lead time, avoid waste, and ensure products and services flow to the customer as efficiently (first pass yield) as possible. Flow is a state that every organization seeks to achieve. If this state is achieved, it can eliminate the need for having a Kanban System (an inventory control system commonly used in just-in-time manufacturing). A VSM always consists of information, process, and time flow. Changes from current state to future state are based on cycle time primarily.

In order for VSMs to be effective, they must be visible where the organization sees and works to it. VSMs use the same format as an A3 (refers to a European paper size similar to a 11"x17" paper size. It is a one page living document describing a current state process problem, root cause analysis, potential

48

countermeasures towards a future state, and an action plan to carry out the countermeasures). The VSM is comprised of a current state, future state, and detailed implementation plan.

It is important to note that process mapping differs from VSM as it does not include information nor time flow; it simply breaks down process steps. Process mapping is one component of a VSM. **(Figure 5-3)**

PROCESS
FLOW
MAP

| PROCESS 01 | → | PROCESS 02 | → | PROCESS 03 | → | PROCESS 04 | → | PROCESS 05 | → | PROCESS 06 |

There are three states of a VSM:
1. Current State - baseline process before changes are implemented. It is mapped and agreed as-is situation and is an illustration of opportunities and pain points.
2. Future State - agreed and defined improvement state in the short-term (1-3 years). It represents the blueprint for planning and executing the improvements for the pain points.
3. Ideal State - agreed and defined improvements state in the long-term (beyond the 3 years of the future state). In most organizations, the ideal state is not used. Instead, the future state is updated periodically.

In addition, there are three types of VSM:
1. Process Level- station-to-station
2. Factory Level - wall-to-wall
3. Extended Level - end-to-end (Sales through Delivery)

Note: breakthrough / innovative thinking can be included in both the future and ideal states. Breakthrough / innovative thinking is defined as a "do different";

something that the organization has not done in the past or something that the organization has done in the past but failed at it. I would recommend capturing unplanned costs vs. planned costs. Unplanned costs might include premium freight. I would also recommend capturing what is controllable and what is noise. One great example that was provided to me years ago was two billion dollars of energy spent to heat water to dye fabric. How might the company eliminate the cost for heating water and still dye the fabric? This is what breakthrough thinking is about.

It's worth also noting that once the mapping is complete, the most difficult part comes next: creating the Detailed Implementation Plan and executing it. The Detailed Implementation Plan is the list of prioritized actions that must be completed in order to close the gap between the states.

Viewing value streams as systems allows us to strategically utilize the most effective Lean tools such as continuous flow, visual controls, and quick changeovers commonly known as Single-Minute Exchange of Die (SMED).

In order to leverage Industry 4.0 technologies such as automation, the process must be in a state of control. I learned from my sensei to automate only after understanding and simplifying the work that humans are doing. This simplifying begins with seeing the value stream.

Caution Points- Where Companies Can Get This Wrong:

- No emphasis placed on the execution from current state to the future state. This execution is guided by the Detailed Implementation Plan. When this occurs, any postings of current and future state maps become nothing more than posting wall art. In my experience, the number one documented required action on the Detailed Implementation Plan is a need for standard work. If you are not transitioning to creating standard work to control the improvements detailed within the future state, you will never actually achieve the future state. In some organizations where execution was accomplished but lack of a control mechanism to prevent "mission creep" was evident, they experienced no progress. We will discuss standard work in Chapter 6: Control Phase.

- People not adequately trained prior to initiating the VSM process. This results is a lack of understanding, motivation, and buy-in. Training must be a well thought out strategy.

- Loss of focus on the customer or end user.

- The event is not supported from the upper ranks of the organization. The lack of support results in a deceleration of team motivation. In most cases, leaders of the organization may not even be aware of the event.

- The organization is function driven only. VSM cannot be successful if done in silos. The entire organization must be involved, including departments that are less function driven.

- Lack of trust. This can occur simply from the person leading the VSM event. This often occurs when a "end-to-end" event takes place. Certain groups are cautious of what they say or are not as transparent as they could be. This behavior is often because of past poor leadership where information was used against the group in some shape or form.

- KPIs not aligned.

- Attempt to create one VSM for more than one product commodity. You must create one VSM per product.

- KPIs driving the wrong organizational behaviors.

- Team not fully understanding or asking what is value added (VA) work and what is non value added (NVA) work. Team members will have differences in opinion so you must ensure alignment.

- Part of the Detailed Implementation Plan to achieve the future state is reducing headcount.

- Making the mistake of focusing their Lean implementation efforts on specific areas and tools of Lean Manufacturing such as cells and kanbans. These efforts have created isolated areas of improvement that typically do not result in significant results or results that the customer experiences.

- A lack of due diligence preparing for the current state event. The event must be well thought out with an adequate amount of planning. A Pre-Work Session should be conducted approximately four weeks prior to the event. The homework assigned to the team ahead of the event should be to create a "straw man" (a best effort at creating the current state map). This map should represent the Worst of the Worst (WoW) projects or programs that have occurred.

- A lack of participation with the invited participants.

- The facilitator not asking the right questions. Questions drive discussions which drive discovery.

- Not identifying where the process pacemaker is. The pacemaker is the first process that turns the product order specific. It is a tool that informs the employees whether work operations are ahead or behind schedule.

- Not creating a VSM prior to initiating an Assessment / Maturity Model and initiating Strategy to Execution (Strategy Deployment).

- Not understanding what the current and future state value delivering a product or service is.

- Not identifying why the handoffs between processes failed during the current state mapping. They do not answer the "what was the expectation and why wasn't it not accomplished." In addition, they do not answer the "how does each process measure success; what are our internal commitments."

- Silo focused. Not paying attention to one's process output is the next processes' input.

- Going through the value stream mapping motions and not placing an emphasis on time to process, manufacture, or service as a commodity. Time dictates cost.

- Not including value stream maps within project business cases, where applicable.

- Once the current state is completed and prior to developing the future state, not asking the question "What are we bad at and why?"

Note: Use Cases (Application Benefits) located in the Appendix

Tool Box:

• BoB / WoW Template	• Parking Lot List	• Positives & Opportunities List
• Impact & Ease of Implementation Matrix	• 4-Blocker	• Lean Training Material
• Countermeasure Template	• Action Tracker	• Executive Report Template

2. 6S:

6S is defined as a methodology that focuses on workplace safety and efficiency through organization (reduce the time spent looking for items; it's a place for everything and everything in its place mentality), cleanliness, standardization, and sustainment of the results. Innovative thinking increases the effectiveness of 6S. It is a leading indicator for organizational discipline.

Most people reading this book are familiar with 5S:
1. Sort
2. Set In Order
3. Shine
4. Standardize
5. Sustain

I categorized 6S as a Simplify Phase tool as the end result is a simplified (decluttered) and organized environment. If something is out of place or is a potential safety hazard, it will be visually detected.

I began my career within a Japanese electronics manufacturing company where sensei taught me 5S + Safety= 6S. Six S is a pillar of the visual factory and makes abnormalities visible.

For me, 6S has always been primarily about safety. A process with great housekeeping practices results in a lower probability of injuries. To validate my point, compare injury rates within a factory that practice 6S vs. one that does not. I will take it a step further, I have been in a manufacturing factory, comprised of many sub feeder lines providing product to two assembly lines, who practiced 6S. The sub feeder lines that had the higher 6S audit scores had the lower injury rates as compared with the lines with lower audit scores who had higher injury rates.

I provided an example of 6S in the Introduction chapter. I stated that most people think with a Lean mindset naturally. To validate this point, think about a time where you cleaned out your garage, attic, or basement and then think about the process you used… sorted, set in order, shined, standardized, and sustained.

Caution Points- Where Companies Can Get This Wrong:

- Last minute 6S culture. The factory and back offices hastily complete 6S the night prior to or the morning of an audit. One way I could always determine if this was the case was to look in the bathrooms prior to conducting the audit. If the bathrooms were not clean, that would be evidence of a last minute 6S. This also applies to the state of the company's quality management process. Prior to a quality audit, look in the scrap bins or dumpsters as an indication of how well the quality management system is working. In addition, when you are walking the factory floor, listen to the sounds … if you hear hammering or cutting, it is an indication of a failure in product engineering.

- During organizational busy times, 6S drops off; it is no longer a priority.

- Organizations that have implemented a shift-to-shift 6S audit checklist encounter issues when one shift leaves the next shift with a process that has not undergone 6S. In this case, leadership does not enforce the expectations and the program becomes a pencil whipping daily exercise.

- Leadership not promoting 6S as much as they should.

- Safety leadership not driving the correlation between high 6S audit scores and low injury rates.

Note: Use Cases (Application Benefits) located in the Appendix

Tool Box:

| • Red Tags | • 6S Audit Check List | • 6S Innovative Picture Template |

3. Production Preparation Process (3P):

Production Preparation Process (3P) is a tool that focuses on removing the 8 Wastes in the early stages of new plant introductions, new product introductions, product design changes, and changes in demand through a new / innovative way of systematic thinking. It accomplishes this through designing the ideal case process, ensuring that everything comes to the product already assembled, working back from the output to the inputs, and prototyping the process.

The objective of 3P is to build right the first time. Whether it is a new plant, new product, product design changes, or changes in demand, it should be done right the first time. The advantages that 3P provides are:
- Ease of manufacturing
- 1st time yield (process / flow)
- Expense reduction
- Vertical start up and low development cost
- Built in up front safety and quality
- Initiate on time

I categorized 3P as a Simplify Phase tool because simplification of design is built into the end result. The simulation component of 3P is detailed in Chapter VIII: Predictive Phase.

The below 3P framework that I learned from the sensei at Shingijutsu, USA(13) is as follows:

- Bring together the right team.

- Provide 3P and 6S training.

- Determine and understand volume to include schedule detail, target launch date and production ramp.

- Begin with a Fishbone Diagram **(Figure 5-8)** as a means of creating a view of the best process flow using the spine of the diagram with the remainder of the diagram representing the sub-feeder lines feeding the process flow. The advantage of a Fishbone Diagram is it shows how everything is connected. Use the current state Bill Of Materials (BOM) removing the component and assembly drawings. Note: Use colored placeholders for components and assemblies not designed yet.

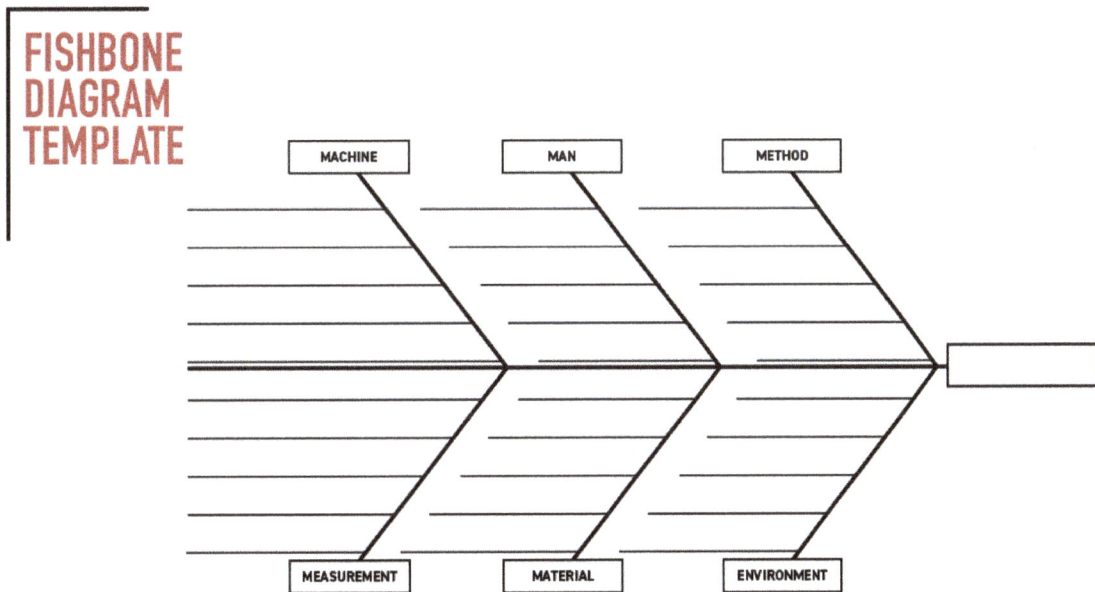

FISHBONE DIAGRAM TEMPLATE

- Complete a Production Capacity Worksheet with a focus on identifying process bottlenecks as defined by units per day at individual processes exceeding planned units per day.

- Build Yamazumi Charts (Japanese word for to stack up) identifying bottlenecks as made visible by process times exceeding takt time. They show the balance of cycle time work loads between a number of operators typically in an assembly line or work cell.

- Create Spaghetti Diagrams detailing travel time.

- Using the Fishbone Diagram, work through a Process at a Glance (3P-7 Ways Tool) form. **(Figure 5-9)** Transfer each block from the Fishbone

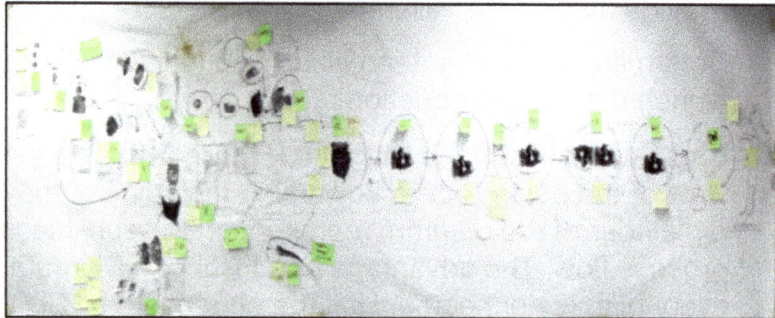

3P - 7 Ways Tool							
(Process at a Glance)							
Assembly _____ Process ____ of _____							
SEQUENCE							
Material Process Sketch							
Process Method							
TIME							
Poka-Yoke No-Go Gauge							
Tools							
Jig/Fixture							
Machine							

3P Production Preparation Process

Name:_____
Date:_____

TAKT Time

Diagram to the Process at a Glance. Sketch what product looks like at the end of each work station using the following categories:

- Material Process sketch will answer the "What"

- Process sketch will answer the "How"

- Poka-Yoke (Japanese word for error-proof) / No-Go Gauge will answer the "Quality Controls"
- Tools will answer the "With what"
- Time will answer the "When"

You should be able to answer whether you have the most efficient way. If not, incorporate the 7 Ways approach using weighed factors. Generate ideas through use of the 7 Ways tool. **(Figure 5-10)** This tool provides a structure to capture 7 ways of doing something such as creating a process layout as an example.

3P - 7 Ways Tool	**3P** Production Preparation Process	Name:_____ Date:_____
		TAKT Time

Assembly _____ Process _____ of _____

SEQUENCE							
Material Process Sketch							
Process Method							
TIME							
Poka-Yoke No-Go Gauge							
Tools							
Jig/Fixture							
Machine							

The steps of doing 7 Ways are as follows:
1. Determine function (what is the theme and scope of the event).
2. Describe ideas (Shingijutsu recommends 7 ideas per person), ideas .should not be discussed with other team members. For example,

team members should not consider machine while working on method. Each idea must be explained either with sketches or verbally.

3. Separate ideas.
4. Sort ideas into groups.
5. Select the three best designs based on criteria.
6. Conduct more Moonshine (rapid prototyping and iteration. It is about creating things from ideas using whatever is available such as cardboard, plastic bottle tops, PVC piping, etc. During the 2D to 3D transitioning within 3P, Moonshine is used to create the 3D) on the top three designs such as building mock-ups of the process method of the three ideas. This step includes conducting experiments and trials to collect real data in order to answer the question "How does each design perform against our criteria?"
7. Gather and evaluate data, such as bottleneck data through the application of Yamazumi Charts for process time estimates and Spaghetti Diagrams for flow, for the top three and select the best one. This step should be accomplished using an Evaluation Checklist. Checklist criteria should be decided by team members who should be thinking and focused on the critical to quality (CTQ) and most importantly, the voice of the customer (VOC). Note: the one selected can be a hybrid of the others.

Next, a 2D representation of the "way" selected is developed. This is accomplished on a table or bench using butcher paper and sketching out the process or product. Then, team members will create a rough bench top mockup from the 2D paper representation. The mockup is accomplished using cups, tape, bottle caps, cardboard, or any available supplies.
Tools for the 2D:
- Takt Time (as defined by the demand of the customer).
- Fishbone Diagram (used to capture the best process flow of the main line and the sub-feeder lines).
- Process at a Glance (captures the Fishbone Diagram information, the sketches of what product looks like at the end of work in station, and answers the "what", "how", "quality controls", and "with what" questions).
- Yamazumi Charts (used to identify where the individual bottlenecks are as compared to take time).

The rough mockup is then transitioned into a small scale mockup. This is normally done on a large space such as a factory floor. The next step is to increase the scale size by transitioning from a small scale to a full scale mockup

(3D). This is accomplished using cardboard, wood, foam, metal, PVC piping, plastic, etc.

Tools for the 3D:
- Moonshine.
- Full scale mockups (3D using Moonshine to create).
- Scale simulations (can be achieved manually and electronically).

Once the 3D mockup has been achieved, work flow simulations should be run. The simulations will identify opportunities that you will use Kaizen to correct. Use Spaghetti Diagrams and Yamazumi Charts (**Figure 5-11 & 5-11b**) to identify

where the bottlenecks may be. Develop, update, and maintain standard work during the simulation process. The purpose of the simulations is to codify design, operator, material flow, information flow, and tooling.

Once the full scale mockup has been completed, enter the work into CAD.

According to Shingijutsu (13), the flow of use 3P tools are:

1. Understand volume
2. Calculate take time
3. Create a Fishbone Diagram
4. Build Production Capacity Worksheet
5. Sketch Assembly Process Chart
6. Conduct 7 Ways on all processes
7. Mock up at scale level
8. Simulate, Kaizen, update Standard Work
9. Repeat Steps 8 multiple times
10. Mockup at full scale level
11. Simulate, Kaizen, update Standard Work
12. Repeat Step 11 multiple times

Caution Points- Where Companies Can Get This Wrong:

- The 3P teams are not creative enough during the 7 Ways, 2D, and 3D steps.
- They throw money at it instead of using the "Moonshine" way of thinking and being creative.
- They don't stick to the 3P framework.
- They do not focus on developing and maintaining the standard work.
- They don't include a team comprised of everyone who will work within the process to do the 7 Ways. More ideas generate the best ways of doing things.
- They don't factor in information and material flow.
- They don't simulate enough.
- They don't adjust the process if something changes such as equipment size.
- Monuments are created such as floor pits and gantry cranes. Monuments prevent the process from quickly (within three days) being changed to a future state.
- Not running pilots in controlled environments. The objective is to not disrupt operation flow. Conducting pilots off line then bringing on line once the pilot has been validated as successful is the ideal situation.

Note: Use Cases (Application Benefits) located in the Appendix

Tool Box:		
• Fishbone Diagram	• Spaghetti Diagram	• Simulation
• Production Capacity Work Sheet	• 3P-7Ways (Process-at-a-Glance)	• Standard Work
• Assembly Process Sheet	• 7 Ways	• Moonshine
• Yamazumi Chart	• Evaluation Check List	• Lean, 3P, and 6S Training Material

4. Disruptive Cost Out (DCO):

Disruptive Cost Out (DCO) is a tool used to drive cost out of an existing product while maintaining the desired levels of product safety, quality, and functionality. DCO is similar to Value Add / Value Engineering (VA/VE).

This tool is categorized in the Simplify Phase as it becomes a forcing function for the organization to analyze their products or services and simplify as a means to drive out cost.

Organizations who are in an aggressive competitive environment where margin erosion is being driven, have to think about and approach costs differently in order to gain a competitive cost position. I have always viewed this as a means to identify the self imposed boundaries.

Caution Points- Where Companies Can Get This Wrong:

- The DCO is not structured nor facilitated correctly.
- The actions required, specifically additional resources, to achieve the cost out actually cost more than the targeted cost per product or service.
- Due diligence in acquiring competitor information does not occur.
- The right people are not invited to be part of the DCO event such as hourly employees, suppliers / vendors, contractors, or former employees.

Note: Use Cases (Application Benefits) located in the Appendix

Tool Box:	
• Business Case	• Cost Out Tracker
• Project Charter	• Obeya Wall or Room

5. Visual Management:

Visual Management is a communication / visual aid tool used to bring awareness to conditions, translates operations status and results, and provide real-time updates. It is used to detect the off-nominal condition immediately such as rubbernecking on a freeway. It should be simple enough that someone external to the company should understand.

This tool is categorized in the Simplify Phase as in order to drive simplification, there needs to be visual management in place to deliver and communicate the changes and end results of the simplification actions. You cannot change what you cannot see.

Jeffrey Liker stated in *The Toyota Way(14),*, *"Humans are naturally visual creatures and are more likely to recall and use information if it is in a visual format, preferably pictures."* A true sign of a Lean organization is one that leverages visual management to make the job of the employees as easy (simplified) as possible. Visualizing work surfaces the opportunities to improve.

Caution Points- Where Companies Can Get This Wrong:

- Asking for "perfect data". They let "I can't get the data" become an excuse instead of figuring out a way to get the data they need.
- Not prototyping use of the Daily Management Boards (DMBs).
- Not ensuring the DMBs drive action or decisions.
- Creating more work managing digital DMBs.
- Measuring the wrong KPIs which drive the wrong behavior.
- Process employees have not been trained on the KPIs with respect to what is being measured and how what they do moves or does not "move the needle".

Note: Use Cases (Application Benefits) located in the Appendix

Tool Box:

- Daily Management Boards
- One Point Lessons
- Digital And Non-Digital Signage
- Light Towers

6. Work Instructions:

Work Instructions are a tool (document) that detail the correct way of successfully completing a task in a step-by-step manner. **(Figure 5-13)** This tool is categorized in the Simplify Phase as in order to document and sustain

Work Instructions Template

KEY POINT REMINDERS — ☐ Critical check — ⊗ Could cause injury — △ Quantity check

Name of Process:		Description/Purpose of Process:
Date of Standard Work:		
Process Owner:		Process Scope:
Needed Tools/Supplies:		

Seq. #:	Process Steps: The "What" Important steps performed that advances the work.	Key Points/Instructions: The "How" *Safety:* Injury avoidance, ergonomics, danger points *Quality:* Defect avoidance, check points, standards *Technique:* Efficient movement, special method *Cost:* Proper use of material	Reasons for Key Points: The "Why" List the reasons for the key points	Training Aid: Diagrams, layouts, screen shots
1				
2				
3				
4				
5				
6				
7				
8				
9				
10				

simplified processes, there needs to be work instructions in place that captures and provides the detail to prevent transitioning back into complexity.

"Without standards, there can be no improvement(15)." Taiichi Ohno

Work instructions normally contain several sheets detailing each step within a specific task. Example: Station 2 is where six wire harnesses are installed into a truck's cab. Work instructions should detail how to install each of the six wire harnesses. Work instructions contain the following elements:

- Process Steps
- Key Points / Instructions
- Reasons for Key Points
- Training Aids (pictures and / or drawings)

Empower team members to create work instructions. When creating work instructions, you should create them in a way that someone off the street walking into the process could read them and generally understand what to do. Naturally, there will be training required. I often have said that someone whose last job was working at a fast food restaurant making sandwiches should be able to come into a process and complete tasks of a technical caliber simply by following the work instructions.

Work instructions are used for new hire training, cross-training, and troubleshooting purposes. They are a living documents that require updating when something in the process changes.

Caution Points- Where Companies Can Get This Wrong:

- Some companies refer to work sequencing sheets as work instructions. Work sequence sheets only detail all the steps in sequential order with the standard time it should take to complete each step. It does not detail how to actually complete each of the tasks. It doesn't explain the "Who", "What", and "Why".
- Work instructions created without the input of the people who are doing the work. Not empowering team members to create work instructions.
- Become divorced from writing work instructions through use of consultants to write them. In most cases, the buy-in piece from the team members working within the process is not there.
- Relying on tribal knowledge and not creating work instructions or relying on tribal knowledge with work instructions in place that does not contain the tribal knowledge. Tribal knowledge typically means that a process has not been evaluated and may not be an optimal process.
- Work instructions created, placed into a binder, and forgotten about.

Note: Use Cases (Application Benefits) located in the Appendix

Tool Box:

- Work Instruction Template

Additional Tools:

- Poke-Yoke- Error Proofing
- One-Point-Lessons (OPL)- Used to highlight a specific point or subject; great for making visible lessons learned as to not repeat the problem. **(Figure 5-14)**
- Kaizen Events (Detail provided in Chapter VII: Proactive Phase)

ONE POINT LESSON

Caution Points- Where Companies Can Get This Wrong:

- Some companies who use Poke-Yoke to mistake proof something (or said another way, idiot proof) inadvertently create idiots. As an example, a company chose to use red, yellow, and green in place of numbers on their process pressure gauges. The process technicians who were responsible for conducting line checks every three hours would simply indicate what color the gauge was indicating at the time of the reading. If you asked the technician what the pressure range should be during a meeting, they would state green and not know that the range was 35-40psi. As another common example, with use of cell phone contacts, how many phone numbers do you actually know?
- Kaizen events should always have a targeted outcome such as a safety / ergonomic risk reduction, cost avoidance, or financial hard savings where the savings flow down through the financial statement. The value is easy to see.

Note: Use Cases (Application Benefits) located in the Appendix

Chapter VI | CONTROL

Control is sustaining results through intentional actions and organizational discipline, which then contributes to and enables an operating rhythm. The objective of control is to prevent "mission creep" or backsliding from the results that have been gained through improvement. Control ensures sustainable, consistent and predictable results. Improvement comes from simplifying, (as discussed in Chapter V: Simplify) being proactive (discussed in the next chapter, Chapter VII: Proactive), and problem solving. In the latter case, controlling the systematic solution(s) assigned to a problem is essential to eliminate the problem and prevent reoccurrence. For this reason, problem solving is included within the Control Phase.

Once control mechanisms are implemented, organizations must monitor them to ensure they are indeed effective. When changes occur through continuous improvement, the control mechanisms will need to be evaluated to ensure they remain effective and if not, they will need to be adjusted to sustain the continuous improvement results.

Sometimes an organization has little to no control over external factors. In these cases, the organization can control their response to those factors. Such a controlled response requires a risk mitigation plan to be created and implemented.

Caution Point for Controlling:
- Be careful to ensure the control mechanisms are well communicated and reviewed with the organization periodically. Remember… if you cannot control it, you don't understand it. This also applies the other way… if you don't understand it, you cannot control it.
- In cases where an organization has little to no control over an event, they can identify the risk and plan to mitigate where applicable. This is where the value is as compared to taking no action.
- Problems that turn into science projects are never resolved and thus controlled.

CONTROL TOOLS

1. Standard Work
2. Lean / Technical Training
3. Key Performance Indicators (KPIs)
4. Business Systems / Operating Models
5. IT Solutions

1. Standard Work:

Standards are a critical part of the foundation of Lean. They ensure that we are not operating businesses on assumptions but rather defining specific requirements. They are an approach to achieve desired outcomes. Standards equal performance targets. Always ask: is performance good enough? Standards are not fixed and should always be improved when process continuous improvement occurs. Continuous improvement begins with standards as they make visible the normal from the abnormal. Any time there is an improvement within the process, a change in machines or tooling, innovation (Industry 4.0), or a staffing increase or decrease, the standard work should be reviewed and updated accordingly.

An example of a standard would be traffic lights that use red, yellow, and green. This standard is recognized globally. Other examples include: regional time zones, identification checks at airports, and flight deck doors installed on all commercial aircraft. Taiichi Ohno stated *"Without standards, there can be no kaizen (improvement.)(15)"*

This tool is categorized in the Control Phase as in order to achieve organizational control, documenting the correct way of completing a task to ensure repeatability and predictability companies will utilize this tool.

There are different levels of standards such as:

- Standard Work (Defines what to do and who should do it. It is important to note that Standard Work is based on time and is correlated to time studies and Yamazumi Charts. This is the difference between Standard Work and Standard Operating Procedures). Standard work is comprised of:
  - Takt time
  - Work sequence
  - Work in process (WIP) / standard inventory
  - Process, material, information layout (I like to include this)

- Job Instruction Breakdown (Used for training purposes):
    - Includes major steps / key points / reasons for key points
    - Used for Training Within Industry (TWI) / teaching; a methodology that provides rapid and consistent training for new hires.
    - Details the "What","How", and "Why"
- Work Instructions- (Defines how to do it):
    - Detailed information of how to complete a task(s).

Standard Work is a methodology; a detailed description of the current best and agreed upon practices for performing a task. Simply put, standard work is everyone doing the same thing, the same way, all the time achieving the same results; a repeatable, predictable outcome. The purpose of standard work is to control the outcome of the process both internally and externally, thus meeting the customers' needs. Standard work ensures gains in improvement are sustained. In the absence of standard work, chaos exists. When we deviate from an established standard, it sticks out; standard work makes variance to the standard visible. If left unchecked, it can lead to the creation of a hidden factory. Any gap from the standard must be addressed through problem solving specifically using the A3 Problem Solving tool which is included within Problem Solving in Chapter VII: Proactive Phase. Once standard work has been implemented, it can improve:
- Flow
- Quality
- Productivity
- Lead Time
- Consistency & Repeatability

Standard work is a tool that illustrates people's interactions as they produce a product or service. It ensures that people, processes, machines, and materials are working in harmony. Standard work must contain what you do, the order in which you do it, and the time required to do it. Standard work is achieved as follows:
- It details the motion of the employee and the work sequence
- It details the best way we currently know and understand (the current state)
- It makes problems visible quicker
- It ensures consistent results

Standard work is a living document to be maintained. It should be based on best current known method that meets internal and external customer needs.

There are three types of standard work:
1. Standard work where repetitive work (minimal variation) is done. In this environment, cycle times do not change for multiple models. Example: automated assembly operation, manual assembly operation.
2. Standard work where where repetitive work (some variation) is done but in this environment, cycle times differ depending on the product type / model. Note: it will require the creations of standard work for each product type / model. Example: An assembly line with a product standard platform but different options, using different machines, tooling, etc.
3. Standard work where repetitive work (variation not always controlled) does not exist; this type is sometimes associated with non-production work such as a job shop environment. Note: it will be necessary to create standard work for each operation. Example: Single machine exchange of dies (SMED), quality check operations, transportation of material and product.

Standard work is comprised of the following management tools:
(Figures 6-1-5)

- Time Observation Sheet
- Standard Production Capacity Sheet
- Standard Work Combination Sheet
- Standard Work Chart
- Yamazumi Chart

- Not taking the time to create standard work.
- Not updating standard work when the process changes.
- Not involving the people doing the work.
- Not linking standard work to the company's Business System.

Note: Use Cases (Application Benefits) located in the Appendix

Tool Box:

• Time Observation Sheet	• Yamazumi Chart	• Job Instruction Breakdown Template
• Standard Production Capacity Sheet	• Standard Work Chart	• Work Instruction Template
• Standard Work Combination Sheet		

2. Lean / Technical Training:

Lean / Technical Training is the process of transferring knowledge to practitioners with the objective of increasing their awareness, understanding, and comprehension.

This tool is categorized in the Control Phase as standard work must be communicated to the people doing the tasks to ensure comprehension. After training the people doing the work, it becomes critical to ensure that standards are followed and maintained. Companies will apply standard work, then will use this tool to ensure their products and services are created and provided as intended.

Training programs are essential for injecting into the company the knowledge that is required to drive Daily Management, Lean Management, and Breakthrough Management. Training programs are also a means of successfully operating the business. Effective training should ensure an engaged workforce. One key component to ensure a successful experience for the trainees is using a qualified and motivating instructor.

Effective training programs are compromised of:

1. Awareness Training: the classroom (in person or virtual) teaching of the material.
2. Practitioner Training: the hands-on application of the awareness training.

Caution Points- Where Companies Can Get This Wrong:

- Training provided by unqualified trainers.
- Too much content such as many tools vs. the basic tools required to begin.
- Making Lean training mandatory. In my experience, this seldom works out and ends up driving the wrong behavior. Offer the training to everyone on a voluntary basis. Business unit leaders can submit employees to the training based on their own rational. It is always in the best interest of the leader to create an army; a force multiplier of problem solvers and problem preventers. The pull is better than the push.
- Companies accepting the "one and done" approach to project work. Once an employee successfully completes a project for certification, they should be excited to use the tools and methodology that they learned to solve or prevent problems they are faced with on a daily basis.

Note: Use Cases (Application Benefits) located in the Appendix

Tool Box:

- Awareness and Practitioner Training Material

3. Key Performance Indicators:

Key Performance Indicators (KPIs) are a quantifiable measure of performance over a period of time. They are used to determine performance against a plan. They should drive a decision or an action. The correct KPI depends on the company's goals and objectives and they should be linked to something such as customer needs. You never know good enough unless you have a target.
This tool is categorized in the Control Phase as KPIs provide an indication whether the business is in a state of control or not. Companies who are focused on understanding whether they are winning or losing utilize this tool.

Attributes of a KPI:
- Measurable
- Available in a timely manner
- Relevant
- Impactful to the financials

Types of KPIs:
- Operational or Process- measures the efficiency, effectiveness, and productivity of an operation or process.
- Input- measures assets and resources used to generate results.
- Output- measures the outcome of business initiatives.

- Leading- measures the input that affects the lagging results. These are real-time results that can be reacted to immediately.
- Lagging- used as a baseline or focus area for improvement.

Note: for purposes of this book, I will focus on leading and lagging indicators where leading indicators are predictors of lagging indicators.

There should be a single point of accountability (SPA) assigned to each KPI whether leading or lagging. It is always a best practice to assign one person verse more than one person. Ideally, these KPIs should be included in the employee's annual performance review.

KPI Trackers or Bowler Charts are tools to track; make visible performance results. These tools, are essentially the same but companies place different labels on them. These trackers or charts measure actual results to planned targets. They can also include previous year's results. I have always viewed these as health reports of a process, operation, machines, etc. Whenever the actual KPI results are not favorable, as indicated in red font or cell, this becomes an opportunity for kaizen and communicating directly with the employees who are closest to the work; the voice of the employee (VOE), as a means to possibly uncover some unknown condition.

Examples of Leading and Lagging Indicators:
(Figure 6-6)

LEADING & LAGGING INDICATORS EXAMPLE		
	LEADING INDICATOR	**LAGGING INDICATOR**
Company	• Quantity of Near Misses Performed • Near Miss Corrective Action Completion Rate • Quantity of 6S Audits • Frequency of Safety Awareness Training • Frequency and Quantity of Safety Audits	**Safety**
	• Pricing • Volume • Cost Reductions	**Contribution Margin**
Personal	• Minutes and Intensity of Cardio Per Day • Minutes and Intensity of High Rep / Low Weight Training Per Week • Smaller / More Frequent Clean Meals	**Weight Loss**

Caution Points- Where Companies Can Get This Wrong:

- Using the wrong KPIs.
- Not understanding the difference between leading and lagging indicators.
- Not prioritizing both Safety and Quality. They lose sight that Safety is a priority and Quality is a standard.
- Using the single point of accountability per KPI as a "throat to choke" verses a single source to better understand current state and what we are doing to improve it.
- Most companies get hung up on Quality and Safety targets being perfect and thus seen as being unrealistic or unachievable. As an example, having a safety Total Recordable Incident Rate (TRIR) annual target of 0 or having a quality Defects Per Unit (DPU) of 0. I have worked within companies who set targets in different ways but I feel the best way is to have a future state and ideal state target where ideal is perfect. Use the

future state target to strive towards the ideal state. This follows the same principle of a value stream map with a current, future, and ideal states.

Note: Use Cases (Application Benefits) located in the Appendix

4. Business Systems / Operating Models:

A Business System, also referred to as a Business Process or Operating System, describes the "how" a business can and should generate value. It leverages the company's strengths, dedicated people, optimized processes, and focused performance to deliver a full potential experience for customers, employees, community, and shareholders. It is a framework of a company's strategy that is aimed at driving and delivering value through execution. The execution of the strategy becomes the formula for success and creating competitive advantage. A Business System is a tool that describes the "how" a business should generate value. It is a framework of a business's vision, mission, values, and strategy that is aimed at driving and delivering value through execution. Note: an online search of Fortune 1000 companies will provide insight into the mechanics of a Business System.

An Operating Model is how the organization is structured in order to deliver value and achieve profitable growth.

These tools are categorized in the Control Phase because they provide the strategic, directional compass and the most optimized organizational structure to achieve the strategic plan. Companies who seek to ensure organizational structure and control over targeted outcomes will use business systems and operating models.

A business system should be guided by the company's vision, mission, and values. Similarly, business systems should be centered on shareholder value generation. This tool is a framework that keeps the organization aligned, prioritized, and focused on winning.

A business system should include the following:
  • A focus on people ("how we work together") , process ("what we work on"), and performance (end result of people + process).

- Strategic Plan (vision / mission / core values)
- Value Creation (for employees / customers / shareholders / community)
- Operational Excellence Capability across all functions

Caution Points- Where Companies Can Get This Wrong:

- Most companies with Business Systems do not leverage them nor use them. They essentially serve two purposes at that point: wall art and provide a feel-good moment for the Wall Street analysts who believe the company is focused, prioritized, and aligned due to being guided by a business system. To validate this point, view a few companies who market a business system within their annual shareholders reports and you will find mention of their business system somewhere within the first four pages or so and then speak with someone who works within the company and you will be shocked as to what you may hear.
- They let their business systems become a religion.
- They don't guard against rigidity.
- Allowing team members to blindly follow and not think about the business system or operating model.
- Changing or eliminating a business system and / or operating model frequently such as a change in leadership results in some employees viewing them as the "flavor of the day".
- Not focusing on the early wins.
- Discounting progress.
- Do not understand the importance of people connect the Business System.
- Not guided by a Business System.
- Not placing energy into the system and it eventually fails.

Note: Use Cases (Application Benefits) located in the Appendix

Tool Box:

- Best Practices (internal & external)

5. Information Technology (IT) Solutions:
IT Solutions will be covered in the Chapter X: Pulling it All Together.

Caution Points- Where Companies Can Get This Wrong:

- Not investing and leveraging IT solutions either internally or externally.
- Not factoring in internal risk mitigation with respect to security when contacting IT solution vendors for solutions. In most cases, if the company IT can provide a suitable substitute, the security risk will be reduced and the internal compatibility with other internal systems may actually be better. Bottom line, check with the company IT Function first.
- Lack of IT Function resources.

Additional Tools:
- 6S (Covered in Chapter V: Simplify)
- Phase Gate (a project management tool also referred to as Toll Gate or Stage Gate)

Chapter VII | PROACTIVE

Proactive is about initiating actions to continuously advance, improve, and innovate. This begins the transformation towards a critical thinking / forward focused organizational culture.

Caution Point for Being Proactive:
Be careful to ensure you are using your resources focused on the areas that will drive value.

PROACTIVE TOOLS
1. Strategy to Execution
2. Problem Solving
3. Assessment / Maturity Model
4. Line / Point Kaizen

1. Strategy to Execution:
The strategy execution framework is called Strategy to Execution (STE). STE is my label for what is commonly referred to as Hoshin Kanri (Japanese word for compass management) or Strategy Deployment. It is a method that closes the gap between business defined strategy and execution through action plans that are executed and measured using KPIs. Hoshin Kanri (strategy) is where "rubber meets air". Execution becomes "rubber meets the road. It is a vehicle that provides a structured approach to execute defined strategy; a process that is focused on transforming defined strategy into realization. Simply stated, it is a methodology with tools that provides organizational focus, priority, and alignment.

It makes visible the organization's "True North," strategic goals and objectives, annual initiatives, KPIs and initiative owners on one sheet of paper referred to as the X-Matrix.
(Figure 7-1)

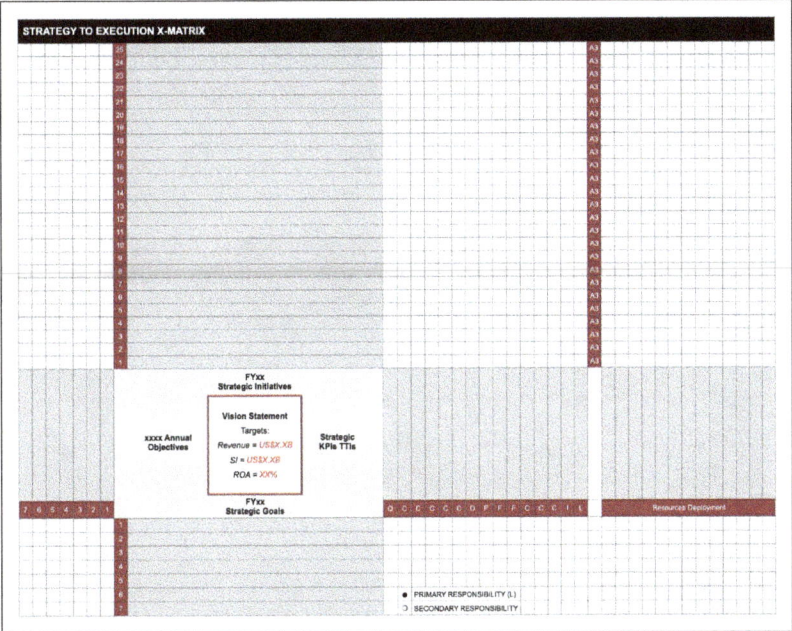

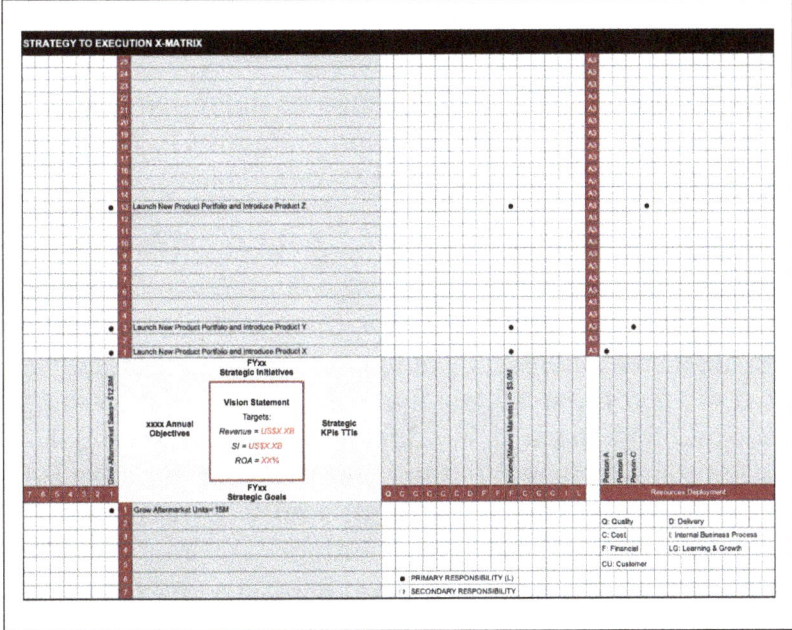

Many organizations find the X-Matrix to be a confusing tool to use. To eliminate the confusion, I created a linear version. The Linear Matrix template can be used instead of the X-Matrix or it can be used to populate the X-Matrix.
(Figure 7-2) By aligning, communicating, and executing defined strategy, STE creates a clear line of sight to success. Success and thus competitive

STRATEGY TO EXECUTION LINEAR MATRIX

	Future State Vision Statement	Future State KPIs	Future State KPI	5-Year Strategic Goals	Annual Objectives	Annual Initiatives	Annual KPIs	Accountability Leader	Priority
	Statement	KPI 1	Select KPI	Goal 1	Objective 1	Initiative 1	KPI 1	Person A	Priority #
		KPI 2				Initiative 2	KPI 1	Person B	Priority #
		KPI 3				Initiative 3	KPI 1	Person C	Priority #
			Select KPI	Goal 2	Objective 2	Initiatives 4	KPI 1	Person D	Priority #
						Initiative 5	KPI 3	Person E	Priority #
					Objective 3	Initiative 6	KPI 1	Person E	Priority #
	Future State Vision Statement	**Future State KPIs**	**Future State KPI**	**5-Year Strategic Goals**	**Annual Objectives**	**Annual Initiatives**	**Annual KPIs**	**Accountability Leader**	**Priority**
Example:	Industry Leader in Aftermarket	Revenue= $1B	Income	Goal 1: Grow Aftermarket Units=15M	Objective 1: Grow Aftermarket Sales= $12.8M	Initiative 1: Launch New Product Portfolio and Introduce Product X	KPI 1: Contribution Margin= $1M	Person A	1
		Income= $120M				Initiative 2: Launch New Product Portfolio and Introduce Product Y	KPI 2: Contribution Margin= $1M	Person B	3
		Market Share= 35%				Initiative 3: Launch New Product Portfolio and Introduce Product Z	KPI 3: Contribution Margin= $1M	Person C	13
Note	KPIs: Key Performance Indicators								
	STE: Strategy to Execution								

advantage is the byproduct of execution. Seneca was quoted at stating *"If a man knows not to which port he sails, no wind is favorable."*

This tool is categorized in the Proactive Phase as most companies see the importance but never commit to the time and resources to actually execute defined strategy. The companies that are forward thinking, will utilize this tool.

According to an article in the Harvard Business Review (December 30, 2015) *Only 8% of Leaders Are Good at Both Strategy and Execution* written by Paul Leinwand, Cesare Mainardi, and Art Kleiner, there is evidence that the majority of global companies are not effective at executing defined strategy(16). The key takeaway is that most companies feel that they are good at creating strategy but not so good at executing strategy. In addition, in 2017 Donald Sull MIT Sloan School of Management Professor of the course *Closing the Gap Between Strategy and Execution* stated that "A recent survey of more than 400 global CEOs revealed that the ability to execute strategy was their number one challenge, ahead of innovation, geopolitical instability, and top-line growth. Executives are right to be concerned. At least two-thirds of large organizations struggle to implement their strategies(17)."

Companies are ineffective at executing strategy for several reasons. First, companies with poor leadership are unable to implement strategies because they lack an execution framework. This may also be due to a lack of buy-in within the organization's leadership. Secondly, they don't know what they don't know. Without the knowledge or resources within the company, they cannot implement a strategy. Third, a lack of discipline means that companies are unable to focus on a strategy execution framework over the needed three to five year strategy planning cycle. Lastly, a strategy could be viewed as additional work on an already resource-constrained organization.

STE methodology lifecycle follows four phases: Plan, Do, Check, Act commonly know as PDCA. **(Figure 7-3)** PDCA changes the process. Once PDCA has been complete, it cycles into Standardize, Do, Check, Act (SDCA). SDCA ensures the change and that the new standard will produce results.

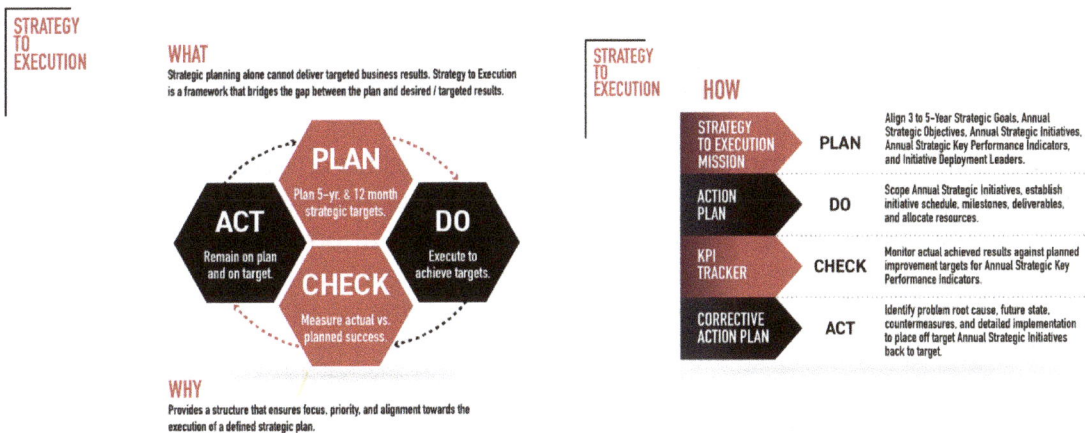

Plan (STE Mission): align 5-year strategic goals (long-term "What"), 1-year strategic objectives (short-term "What"), annual initiatives (the "How"), initiative owners, and target results (KPIs)
Do (STE Action Plan): scope annual initiatives; set initiative schedule, milestones, and deliverables; and allocate resources utilization
Check (STE KPI Tracker): monitor actual results against planned improvement targets Act (Corrective Action Plan): identify problem's root cause, desired state, countermeasures, and remedy actions to transition to favorable results

EXAMPLE OF HOW IT WORKS:

PLAN

The plan phase starts with ensuring that there is a defined enterprise strategy in place. This can be achieved by viewing the enterprise STE X-Matrix. The objective is to ensure a cascading down from the enterprise down to the divisions. If there isn't a VSM for each of the divisions processes, complete a current state VSM to reveal potential opportunities. Conduct a Strengths, Weaknesses, Opportunities, and Threats (SWOT) Analysis followed by executive interviews. Then identify the True North statement and targets such as Revenue, Segment Income, and Return on Assets (ROA). Note: these will be where the division seeks to be in 5 years. Next identify what the 5-year goals are that will ensure ~70% (~30% is delivered through Daily and Lean Management) of the True North is achieved within the 5-year time period. Note: ideal quantity of goals should be between 5 and 7. This will be followed by identifying what the 1-year objective(s) are per 5-year goal. Note: the ideal quantity of objectives per goal is 1 to 3. Identify what the annual initiative(s) are per 1-year objectives.

Important notes to keep in mind are that there is is no rule of thumb for how many initiatives you should have per 1-year objective but ensure you have the resource capacity to complete. The initiatives should be a "do different; breakthrough" For example, an initiative could include something you are not sure how you will accomplish or something that was attempted in the past with no success. I have heard it mentioned to think with a *Back to the Future* mindset. These cannot be "Run the Business" initiatives. Ideation sessions or as I like to refer to them, "In the lab sessions" are the best method to generate ideas. The mindset should be ideation through execution.

STE years 1 through 3, use larger breakout teams. Years 4 and 5, use smaller breakout teams. Within a business dynamic state, breakthrough initiatives should be easier to identify as compared to a static state due to opportunities when the business is operating.

Action Plan created per annual initiative using a Action Plan tool **(Figure 7-4)** is comprised of initiative scope, plan, deliverables, and resources. Action Plans and targets are created by the process of cross-functional collaboration known as "catchball"

The KPI Tracker Chart **(Figure 7-5)** is created per annual initiative and is comprised of planned target and actual results by month. The Annual Action Plan is then finalized with the planned completion date. The Annual Strategic KPI Tracker is finalized with a planned target and completion date.

STRATEGIC KPI TRACKER CHART									Reporting Month & Year:	XX	XXXX

Annual Strategic Initiative Name:		Strategic KPI Tracker - Strategy to Execution

Annual Strategic Initiative Key Performance Indicators (KPIs)	Strategic KPI Impacted	UOM	CDC	YE Target	Trend	YTD		Oct-14	Nov-14	Dec-14	Jan-15	Feb-15	Mar-15	Apr-15	May-15	Jun-15	Jul-15	Aug-15	Sep-15
1						0	Plan / Actual												
2						0	Plan / Actual												
3						0	Plan / Actual												
4						0	Plan / Actual												
5						0	Plan / Actual												
6						0	Plan / Actual												
7						0	Plan / Actual												
8						0	Plan / Actual												
9						0	Plan / Actual												
10						0	Plan / Actual												
11						0	Plan / Actual												
12						0	Plan / Actual												

Strategic Key Performance Indicators (KPIs)	B/C	UOM	CDC	YE Target	Trend	YTD ACT		Oct-14	Nov-14	Dec-14	Jan-15	Feb-15	Mar-15	Apr-15	May-15	Jun-15	Jul-15	Aug-15	Sep-15
1						0	Plan / Actual												
2						0	Plan / Actual												
3						0	Plan / Actual												
4						0	Plan / Actual												
5						0	Plan / Actual												
6						0	Plan / Actual												
7						0	Plan / Actual												
8						0	Plan / Actual												
Summary Benefits						0	Plan / Actual												
Summary Costs						0	Plan / Actual												

Next the annual STE Resource Utilization Analysis **(Figure 7-6)** is reviewed and agreed upon across all initiatives to avoid over-utilization. It is possible to have functional people supporting multiple initiatives so ensure that their individual resource utilization accounts for all the initiatives they are collectively

M = Month I Y = Year

Annual Strategic Initiative	Resource Name	M/Y	M/Y	M/Y	M/Y	M/Y	M/Y	M/Y	M/Y	M/Y	M/Y	M/Y	M/Y	M/Y	M/Y	M/Y
		0%	0%	0%	0%	0%	0%	0%	0%	0%	0%	0%	0%	0%	0%	0%
		0%	0%	0%	0%	0%	0%	0%	0%	0%	0%	0%	0%	0%	0%	0%
		0%	0%	0%	0%	0%	0%	0%	0%	0%	0%	0%	0%	0%	0%	0%
		0%	0%	0%	0%	0%	0%	0%	0%	0%	0%	0%	0%	0%	0%	0%
		0%	0%	0%	0%	0%	0%	0%	0%	0%	0%	0%	0%	0%	0%	0%
		0%	0%	0%	0%	0%	0%	0%	0%	0%	0%	0%	0%	0%	0%	0%
		0%	0%	0%	0%	0%	0%	0%	0%	0%	0%	0%	0%	0%	0%	0%
		0%	0%	0%	0%	0%	0%	0%	0%	0%	0%	0%	0%	0%	0%	0%
		0%	0%	0%	0%	0%	0%	0%	0%	0%	0%	0%	0%	0%	0%	0%
		0%	0%	0%	0%	0%	0%	0%	0%	0%	0%	0%	0%	0%	0%	0%
		0%	0%	0%	0%	0%	0%	0%	0%	0%	0%	0%	0%	0%	0%	0%
		0%	0%	0%	0%	0%	0%	0%	0%	0%	0%	0%	0%	0%	0%	0%
		0%	0%	0%	0%	0%	0%	0%	0%	0%	0%	0%	0%	0%	0%	0%
		0%	0%	0%	0%	0%	0%	0%	0%	0%	0%	0%	0%	0%	0%	0%
		0%	0%	0%	0%	0%	0%	0%	0%	0%	0%	0%	0%	0%	0%	0%
		0%	0%	0%	0%	0%	0%	0%	0%	0%	0%	0%	0%	0%	0%	0%
		0%	0%	0%	0%	0%	0%	0%	0%	0%	0%	0%	0%	0%	0%	0%
		0%	0%	0%	0%	0%	0%	0%	0%	0%	0%	0%	0%	0%	0%	0%
		0%	0%	0%	0%	0%	0%	0%	0%	0%	0%	0%	0%	0%	0%	0%
		0%	0%	0%	0%	0%	0%	0%	0%	0%	0%	0%	0%	0%	0%	0%
		0%	0%	0%	0%	0%	0%	0%	0%	0%	0%	0%	0%	0%	0%	0%
		0%	0%	0%	0%	0%	0%	0%	0%	0%	0%	0%	0%	0%	0%	0%
		0%	0%	0%	0%	0%	0%	0%	0%	0%	0%	0%	0%	0%	0%	0%
		0%	0%	0%	0%	0%	0%	0%	0%	0%	0%	0%	0%	0%	0%	0%
		0%	0%	0%	0%	0%	0%	0%	0%	0%	0%	0%	0%	0%	0%	0%
		0%	0%	0%	0%	0%	0%	0%	0%	0%	0%	0%	0%	0%	0%	0%
		0%	0%	0%	0%	0%	0%	0%	0%	0%	0%	0%	0%	0%	0%	0%
		0%	0%	0%	0%	0%	0%	0%	0%	0%	0%	0%	0%	0%	0%	0%
		0%	0%	0%	0%	0%	0%	0%	0%	0%	0%	0%	0%	0%	0%	0%
		0%	0%	0%	0%	0%	0%	0%	0%	0%	0%	0%	0%	0%	0%	0%
		0%	0%	0%	0%	0%	0%	0%	0%	0%	0%	0%	0%	0%	0%	0%
		0%	0%	0%	0%	0%	0%	0%	0%	0%	0%	0%	0%	0%	0%	0%
		0%	0%	0%	0%	0%	0%	0%	0%	0%	0%	0%	0%	0%	0%	0%

supporting. In addition, time allocated for support people doing the roles that they were hired for will need to be accounted for. With all of the above actions completed, the Annual STE X-Matrix or Linear Matrix can be completed after all initiatives have been prioritized and the Annual Strategic KPIs are thus finalized.

DO

- Teams complete the actions as detailed in the Action Plans milestones.

CHECK

- Review the STE Dashboard **(Figure 7-7)** with a focus on the initiatives that have "red" status monthly results.

Strategy To Execution Dashboard			Business Name			Date						Explanation	Last Updated
Strategic Priority Initiatives	Priority	Deployment Leader	Project Owner	Estimated Business Value $,000	Overall	Value Assessment							
						Schedule	Benefits	Costs	Resources	Change Commitment			
STE Initiative Name		[Name]		$0									
STE Initiative Name		[Name]		$0									
STE Initiative Name		[Name]		$0									
STE Initiative Name		[Name]		$0									
STE Initiative Name		[Name]		$0									
STE Initiative Name		[Name]		$0									
Summary of Total Business Value				$0									

On target ☐ Risk of missing target(s) ☐ Target(s) missed ☐

Value Assessment Guideline: 1 Red risk = Red risk overall
3 Yellow risks = Red risk overall

- Initiative owner provides a deep dive review of the initiative Action Plan and Strategic KPI Tracker.

ADJUST

- Initiative owner and team completes a Corrective Action A3 **(Figure 7-8)** for any initiatives that have "red" status monthly results.

Focus: **Owner:**   **Date:**

1. Problem (Business Case- (Why are We Here?)

5. Proposed Countermeasures (What are your proposed actions to close the gap between current / future state?)

2. Current Conditions (Current Status- Where do we Stand?)

This Year's actual performance is X

This Years target performance is Y

3. Goals / Targets (What specific outcomes are expected?)

4. Analysis (What are the root cause(s) of the problem?)

Pareto chart of root causes

6. Plan (What activities are required for implementation and who is responsible for what and when?)

Start Date	End Date	Oct-13	Nov-13	Dec-13	Jan-14	Feb-14	Mar-14	Apr-14	May-14	Jun-14	Jul-14	Aug-14	Sep-14	Status	Impacted Strategic KPI	Effect of Impact (Quantitative)
01-Oct	05-Oct													Complete		
06-Oct	12-Oct													Complete		
13-Oct	30-Oct													Complete		
01-Nov	14-Nov													Complete		
15-Nov	28-Nov													Complete		
01-Dec	07-Dec													Complete		
08-Dec	20-Dec													Complete		
02-Jan	20-Jan													Complete		
21-Jan	30-Jan													Complete		
01-Feb	10-Feb													Complete		
10-Feb	24-Feb													Complete		
25-Feb	30-Mar													Complete		
01-Apr	30-Jun													On Plan	OE Revenue	> USD $5M
02-May	15-Jul													Past Due		
01-Jul	30-Sep													At Risk	OE Revenue	> USD $50M
20-Sep	30-Sep													On Plan	AM Revenue	> USD $70M

7. Follow-Up (What issues can be predicted?)

PLAN (ANNUAL REFRESH)

- The STE X-Matrix or Linear Matrix is reviewed with the objective of determining if anything has changed with respect to the 5-Year Goals, 1-Year Objectives, or Annual Initiatives. Note: unless something has changed within the external environment, the 5-Year Goals should not change. The 1-Year Objectives do sometime change and the Annual Initiatives can change due to the following reasons:
  - The initiative has been successfully completed
  - The initiative is no longer applicable and thus can be removed
  - The initiative is foundational for the current year and thus will be rolled into the following year

- The initiative began after the first of the year and thus will roll over into the new year. Annual Initiatives remain active until their one year anniversary
- A new initiative has been added

Tools used within the STE methodology are:
- Executive Interview Check Sheet
- SWOT Analysis
- VSM
- Impact & Ease of Implementation Matrix
- X-Matrix or Linear Matrix
- Action Plan
- Resource Utilization Analysis Tracker
- Strategic KPI Tracker
- Corrective Action A3
- STE Dashboard
- Report Out Template
- Action Register
- Lesson Learned Repository
- Future Focused STE Initiative Parking Lot

Alan Lakein stated that "Planning is bringing the future into the present so you can do something about it now(18)." If we follow the Covey Quadrants **(Figure 7-9)**, STE is focused on the Not Urgent but Important quadrant. The Urgent and

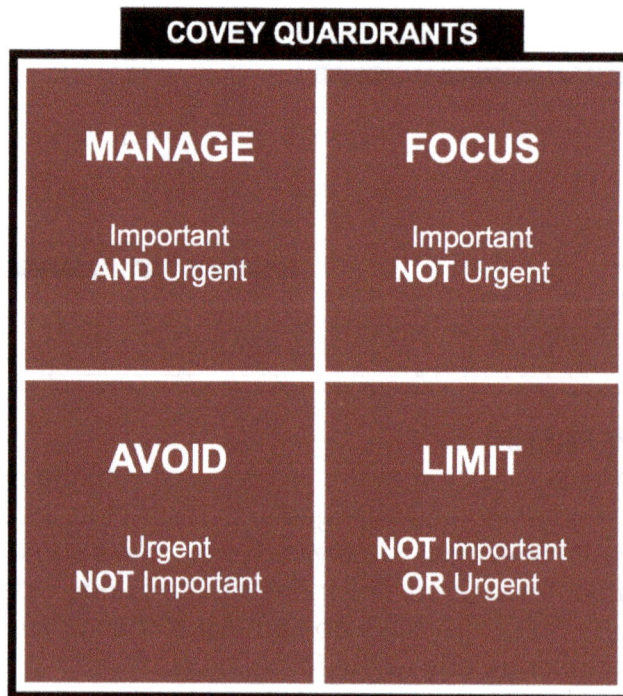

COVEY QUARDRANTS

MANAGE Important **AND** Urgent	FOCUS Important **NOT** Urgent
AVOID Urgent **NOT** Important	LIMIT **NOT** Important **OR** Urgent

Important quadrant is where Daily Management with Problem Solving (Lean Management) applies. This is because STE is about the 3 to 5-year journey of executing the strategic plan. With STE, direction is much more important than speed.

STE is categorized within Breakthrough Management whereas run rate, "run the business" / routine work" is categorized within Daily Management. Improving daily work through Kaizen is categorized within Lean Management. This management system is referred to as the Execution Management System. **(Figure 7-10)** Lean initiatives do not consume resources as breakthrough annual initiatives do. This cycle is a closed loop system:

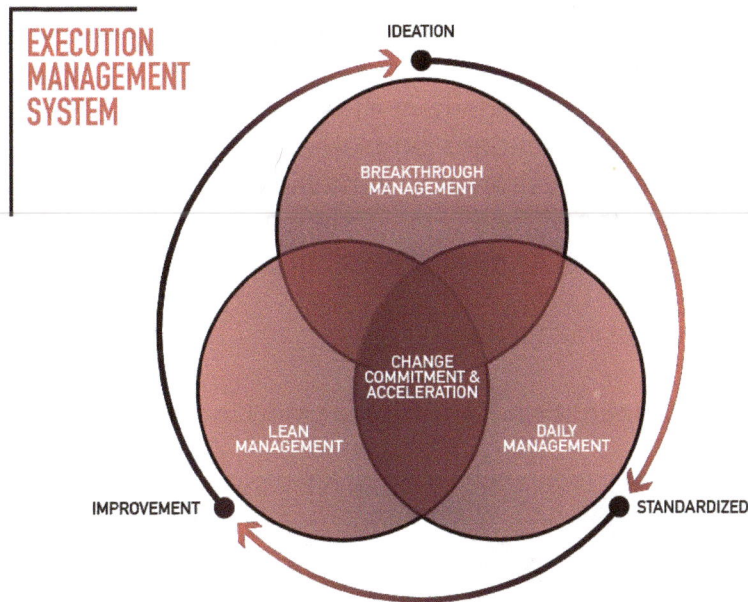

- STE Annual Initiatives that have been completed transition from Breakthrough Management into Daily Management when the initiative has been completed and standardized through standard work (Plan-Do-Check-Act to Standardized-Do-Check-Act)
- From Daily Management to Lean Management when there is a need for improvement or a problem to solve.
- From Lean Management to Breakthrough Management when an innovative approach is required to improve further.

Caution Points- Where Companies Can Get This Wrong:

- No clearly defined strategy.
- Not first focusing on running the business well prior to focusing on adding STE.
- Not focused on mobilizing true commitment but rather engage in "lip service".

- Failure to consider the extent of organizational culture change that is required. It has to feel like a transformation and not just another project.
- Lose focus on the "why" and thus lose their way which drives compliance and not commitment required to achieve success.
- Lack of "catch-ball" across the organization.
- STE annual initiatives that require IT enablement sometimes the enablement is not known until the initiative is in progress. When this fact is discovered, these initiatives, most of the time, fall out of focus due to a lack of IT resources.
- Not divorcing Daily Management KPIs from STE objectives.
- Not adhering to the standard templates.
- Same annual initiatives reported in multiple meetings.
- Not adhering to the agreed upon cadence.
- Duplicate annual initiatives as a result of same initiative but worded differently.
- Too many annual initiatives.

Note: Use Cases (Application Benefits) located in the Appendix

Tool Box:

- X-Matrix or Linear Matrix
- Action Plan
- KPI Tracker
- Corrective Action A3
- Lessons Learned Repository

- STE Initiative Dashboard
- Plan Dashboard
- SWOT Analysis
- VSM

- Executive Interview Check List
- Action Register
- Report Out Template
- Future Focused Idea List

2. Problem Solving:

Before we define problem solving, let's define what a problem is. A problem is simply a situation where the "is" does not equal the desired "want." I heard it once said that all problems are solvable except the problem of losing your mind. This could be, in fact, true but I recommend understanding if a specific problem is under your control (internal) or outside of your control (external). This thinking narrows the problem solving process down. What you cannot control will need evaluated to determine if you can influence the outcome.

Problem Solving is the process of identifying a problem using a set of tools to define, measure, analyze, implement improvements, and control the results. The problem solving process remains the same regardless of the complexity of the problem but the complexity of the tools must increase. It is worth noting that the complexity of the problems we face sometimes dictates new approaches and ways of thinking. This is the kind of innovative thinking that I thrive on and hopefully after implementing the 4 Phase Approach Model, you will too.

This tool is categorized in the Proactive Phase because even though all companies face problems on a daily basis, most companies do not operate with a standardized, structured problem solving framework. In addition, these companies lack the knowledge of how to correctly solve problems. The companies that are committed to solving problems with the mindset to prevent reoccurrence, will utilize this tool.

The role that Lean and Six Sigma plays within problem solving:
- Lean is a philosophy, a way of thinking guided by principles comprised of tools. Lean is focused on placing a process within a state of control through a focus on continuous improvement.
- Six Sigma is simply a set of tools used to reduce or eliminate variation and defects. For Six Sigma to be effective, the process must be in a state of control. For this reason, Lean and Six Sigma merged to be Lean-Six Sigma.

I feel there is a need to make the delineation between the two.

My experience has shown me often people focus on the fallacies of Six Sigma and how it is not what it has been packaged to be. My approach has always been not to focus on that but rather to use the problem solving methodologies contained under the Six Sigma umbrella such as:

DMAIC- Define | Measure | Analyze | Improve | Control
DMADV- Define | Measure | Analyze | Design | Verify
IDDOV- Identify | Define | Develop | Optimize | Verify

I have used all of the above methodologies with success. At the end of the day, where the value resides is in the ability to identify and address the root cause of a problem with the objective of preventing it from reoccurring. Don't get caught up in the labels but rather use the problem solving methodologies.

Knowing which methodology to use for a specific problem is key. Provided below is a Decision Tree that acts as a guide helping you to match the most effective tool to a problem:

(Figure 7-12)

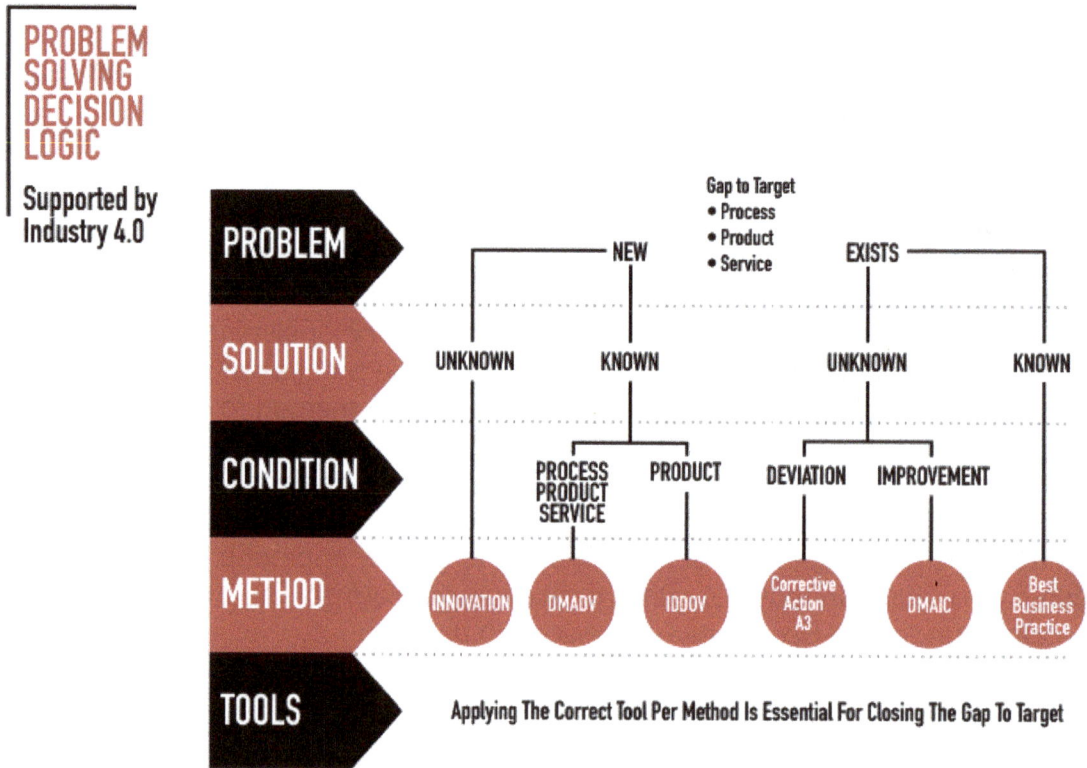

PROBLEM SOLVING DECISION LOGIC

Supported by Industry 4.0

PROBLEM — NEW — EXISTS

Gap to Target
• Process
• Product
• Service

SOLUTION — UNKNOWN — KNOWN — UNKNOWN — KNOWN

CONDITION — PROCESS PRODUCT SERVICE — PRODUCT — DEVIATION — IMPROVEMENT

METHOD — INNOVATION — DMADV — IDDOV — Corrective Action A3 — DMAIC — Best Business Practice

TOOLS — Applying The Correct Tool Per Method Is Essential For Closing The Gap To Target

Caution Points- Where Companies Can Get This Wrong:

- Not being wary of solutions looking for problems.
- Not understanding causes which means they won't be addressed.
- Fail to build a root cause problem solving culture.
- Companies avoiding or not driving using problem solving methodologies such as Define, Measure, Analyze, Improve, Control (DMAIC). The DMAIC framework is a change or shifting in how companies should think about problem solving.
- Problem-focus is not viewed as more advantageous than solution-focus.
- Leaders developing the plan in isolation without the collaboration of their teams.
- Assuming you know the problem you are trying to solve.

- Starting with solutions.
- Not ensuring that strong control measures are in place to ensure sustainability of the improvements.

Note: Use Cases (Application Benefits) located in the Appendix

3. Assessment / Maturity Model:

An Assessment / Maturity Model is a tool used to audit defined criteria and identify the level of maturity a process is currently at with the objective of improving maturity towards a future state. They are used to ensure the organization is working on the right things. **(Figure 7-13)**

	ELEMENT	LEVEL 1 Little to No Action	LEVEL 2 Developing	LEVEL 3 Effective	LEVEL 4 Excellent	LEVEL 5 Max-Potential	Element Score		Weighted Score
1.	PILLAR						0.0	What does a Level 5 Equate to: Value ($)	0.0
1.1	Element #1	Behavior that represents the level	Behavior that represents the level	Behavior that represents the level	Behavior that represents the level	Behavior that represents the level	0.0		
1.2	Element #2	Behavior that represents the level	Behavior that represents the level	Behavior that represents the level	Behavior that represents the level	Behavior that represents the level	0.0		
2.							0.0	What does a Level 5 Equate to: Value ($)	0.0
							0.0		
							0.0		
3.							0.0	What does a Level 5 Mean- Value ($)	0.0
							0.0		
							0.0		
4.							0.0	What does a Level 5 Mean- Value ($)	0.0
							0.0		
							0.0		

This tool is categorized in the Proactive Phase as many companies audit their processes but fall short of developing actions plans to advance audit criteria maturity. The companies that are proactive and seek to drive annual incremental value through a focus on maturity across the organization's value stream will utilize this tool.

An audit is a systematic, independent and documented process for collecting evidence and evaluating it objectively to determine the extent to which audit criteria are achieved. It is validation that you are doing what you say you will do. It establishes a consistent criteria to measure processes and highlight continual improvement. In addition, it leverages internal and external best practices that can then be shared across the company.

Audits are performed for two basic reasons:
1. To determine if requirements are being met
2. To determine where there are opportunities for improvement

The assessment component of the Model is comprised of pillars and elements. As an example, if we were creating an Assessment / Maturity Model for a Maintenance Function, some of the pillars would be:
- Maintenance Strategy
- Maintenance Planning
- Planned / Scheduled Maintenance
- Predictive Maintenance

An example of some of the elements that would fall under a pillar such as Predictive Maintenance would be:
- Identification
- Schedule Compliance
- Condition Monitoring
- Maintenance Standards
- Execution of Maintenance

The maturity component of the Model is comprised of maturity levels and behaviors. Maturity levels can be labeled in many different ways with different quantities of levels. I normally use the following:
- Level 1- Not Meeting
- Level 2- Developing
- Level 3- Effective
- Level 4- Excellent
- Level 5- Max-Potential

Note: you can use just the first three levels if you choose and change the labels but ensure that the label corresponds to the increasing maturity number.

Maturity behaviors are the current state associated with each of the maturity levels. As an example, under the Predictive Maintenance pillar, the Schedule Compliance element the behaviors would look like this:

Level 1 - No planned predictive maintenance performed.
Level 2 - Predictive schedules are developed and compliance is >75%.
Level 3 - Predictive work is scheduled automatically and is executed on time >90% of the time.
Level 4 - Predictive maintenance schedule compliance is monitored and reported monthly with compliance >95%.
Level 5 - Predictive maintenance schedule compliance is >98%. Schedules are evaluated periodically and modified based on equipment performance and importance / priority.

Looking at the example model **(Figure 7-14)**, the objective each year is to increase in maturity level per element. With the behaviors per maturity level being identified, this provides the "How" to achieve the annual improvement.

ELEMENT		LEVEL 1 Little to No Action	LEVEL 2 Developing	LEVEL 3 Effective	LEVEL 4 Excellent	LEVEL 5 Max-Potential	Element Score		Weighted Score
1. Supplier Relationship Management (SRM)							0.0	What does a Level 5 Equate to: Value ($)	0.0
1.1	Vendor Scorecard	Vendors are transactional; not measuring.	Identifying scorecard metrics.	Scorecard developed and in place for all Suppliers / vendors.	Metrics developed and adjusted with vendor.	Adjusting business practices based on data with leading and lagging indicators (KPIs) included.	0.0	Level 5 achievement: $1.5M	
1.2	Shared Opportunities	Transactions are location based.	Transactions are regional based.	Global initiatives are considered.	Transactions are global.	Global alignment around the 5-Year Capex.	0.0	Level 5 achievement: $280K	
1.3	Vendor Shared Opportunities	Vendor fills orders.	Business reviews SKU spend.	Vendor reviews SKU spend and provides consolidation opportunities.	Vendor reviews SKU spend and provides price reduction opportunities.	Vendor analyzes spend and suggests areas for quality/cost improvements.	0.0	Level 5 achievement: $1.1M	
1.4	Relationship Development	Vendor works on a P.O. basis.	Vendors engaged in medium to long term contracts.	Vendor viewed as a strategic business partner.	Vendor provides information on market changes as they happen.	Vendor provides insight and forecasting helping to drive the budgeting process.	0.0	Level 5 achievement: $756K	
2. Reporting & Documentation							0.0	What does a Level 5 Equate to: Value ($)	0.0
2.1	KPIs	Little to no evidence of KPI usage.	Alignment with stakeholders to include usage of correct KPIs (ex. OTD, quality, etc.).	Evidence of KPI implementation and score carding.	Evidence of KPI implementation, score carding and scheduled reviews.	Scorecards include leading indicators.	0.0	Level 5 achievement: $83K	
2.2	Meeting Cadence	Little to no evidence of reviews / reporting.	Evidence of reviews / reporting occurring but not at a consistent basis.	Evidence of reviews / reporting occurring at a consistent basis utilizing a standardized template.	All identified action items have been documented and assigned.	Agenda prepared in advance and meeting is kicked off with progress report from prior meeting.	0.0	Level 5 achievement: Foundational	
2.3	Linkage to Financial Reporting	Little to no evidence of linkage to the Finance Function.	Evidence of linkage to the Finance Function but not on a consistent basis.	Evidence of linkage to the Finance Function on a consistent basis.	Use of cross functional teams leveraging the ideation process.	Utilizing shared metrics for reporting.	0.0	Level 5 achievement: Foundational	
2.4	Contract Management	Contracts not tracked in any database nor shared on a network drive.	Some evidence of contracts being tracked in a database or shared network drive but not on a consistent basis.	Contracts uploaded to a shared network drive and tracked on a consistent basis.	Contracts uploaded to a shared network drive. Contract prices and rates tracked through CMMS.	Contracts uploaded to a shared network drive. Contract prices and rates tracked through CMMS. Vendor compliance to terms & conditions tracked (life cycle management).	0.0	Level 5 achievement: $1.5M	
3. Strategy							0.0	What does a Level 5 Equate to: Value ($)	0.0
3.1	Planning	No evidence of linkage to the 5-Year Strategy Plan.	Evidence of some linkage to the 5- Year Strategy Plan.	Linkage to the 5-Year Strategy Plan.	Strategy plan developed based off of the linkage to the 5-Year Strategy Plan.	Evidence of strategy execution to the defined Procurement strategy plan.	0.0	Level 5 achievement: $210K	
3.2	Supplier Research	Little to no evidence of supplier research.	Market information collected.	Pool of vendors, beyond preferred, determined.	Pool of vendors analyzed.	Analysis used to determine primary vendor.	0.0	Level 5 achievement: Foundational	
3.3	Innovation	No evidence of use of innovation to drive organizational competitive advantage.	Areas of innovation opportunities identified.	Some evidence of leveraging the innovative process.	Perform risk / reward analysis on innovative opportunities.	Evidence of Use of innovation to drive organizational competitive advantage.	0.0	Level 5 achievement: $1.0+M	
4. Sourcing							0.0	What does a Level 5 Mean-Value ($)	0.0
4.1	Standard Operating Procedures (SOPs)	Little to no evidence of identified SOPs.	SOPs identified but not documented.	Use of SOPs on a consistent basis utilizing standardized template.	SOPs are controlled within the Procurement function.	SOPs consistently reviewed and improved by the end users.	0.0	Level 5 achievement: Foundational	
4.2	Demand & Options	Little to no evidence of parts research.	Alternate parts viability information collected.	Alternate parts, beyond preferred, determined.	Alternate parts analyzed.	Analysis used to determine alternate parts and data captured in CMMS.	0.0	Level 5 achievement: Foundational	

From an auditing perspective, I recommend conducting a self-assessment in addition to a peer assessment.

Benefits of implementing an Assessment / Maturity Model are:
- Identifies, shares and operationalizes the best ways of doing things across the business with streamlined and common standards.
- Creates collaboration and increased knowledge of each business.
- Provides a framework for keeping the focus on the elements that lead to productivity outside those factors that can't be controlled.
- Improve internal and external customer relationships.

Caution Points- Where Companies Can Get This Wrong:

- Comparing Assessment / Maturity Model scores between businesses. Friendly competition is great but ensure an apple is being compared against an apple and the wrong behavior is not being driven.
- The Assessment / Maturity Models are too long with audit items.
- The Assessment / Maturity Model contains too many maturity levels.
- The Assessment / Maturity Model is not tied to annual performance reviews.
- Getting stuck in the piloting mode and never transition to other processes.

Note: Use Cases (Application Benefits) located in the Appendix

Tool Box:		
• Assessment / Maturity Model Template	• Leading & Lagging KPIs	• Action Tracker

4. Line & Point Kaizen:

The term Kaizen is a Japanese word that translates to change for the better or more commonly referred to as continuous improvement. Organization's should always be focused on Kaizen. Kaizen should be done each day. Kaizen is mainly applied when there is an improvement opportunity or problem to resolve within Daily Management. This is where Lean Management is applied. Kaizen events target prioritized areas for improvement where rapid improvement can be achieved.

These tools are categorized in the Proactive Phase as most companies do not take the time to conduct Kaizens. The companies that seek continual, continuous improvement will utilize these tools.

Line Kaizen is an improvement that is targeting a line or process. Target time from initiation to completion is within a 3-day period.

Point Kaizen is an improvement that is smaller in nature such as 6S'ing an area within a cell. Target time from initiation to completion is within a 24-hour period.

Kaizen should always be a mindset and Kaizen events should not be viewed as just a tool or a project. Kaizen should be leveraged each day. Kaizens must be fast paced focused on incremental improvement.

Selected areas for Kaizen events should have a Kaizen A3 completed and presented for buy-in, selection, and support. **(Figure 7-15)**

Once selected, and after the Kaizen, actions identified to achieve the future / ideal state should be documented on a Kaizen Detailed Implementation Plan for tracking purposes. **(Figure 7-15b)**

KAIZEN DETAILED IMPLEMENTATION PLAN	EVENT NAME		EVENT NUMBER	Date	
				TAKT TIME	
ACTION ITEM	OWNER		DUE DATE	STATUS	%

Caution Points- Where Companies Can Get This Wrong:

- The best short-term solution may not be the best long-term solution. Solutions will need to be evaluated.
- Not committing to completing Kaizen Event required action items.
- Not being aware that short-term results make people uncomfortable.
- Not creating a Kaizen mindset culture.

Note: Use Cases (Application Benefits) located in the Appendix

Tool Box:

• Kaizen A3   • Kaizen A3 Detailed Implementation Plan   • Kaizen Sheet

Additional Tools:

- Cause and Effect Check Sheets- a hybrid tool combining Fishbone Diagrams with Check Sheets.
- Just Do It (JDI)- action that does not require extensive planning and resources to accomplish.
- Obeya Room or Wall- a designated meeting area where daily huddling and planning takes place.
- Moonshine- an approach to problem solving that uses limited resources to drive innovative thinking and end results. An example would be a custom car builder who uses cardboard and masking tape to form a body panel, ensure fit, form, and function prior to committing to cutting & forming metal, and installing the panel.
- 5W1H- a tool that is comprised of questions to ask when a problem first occurs: What, When, Where, Who, Why, and How.
- Fishbone Diagrams- detailed in Chapter III: Simplify.
- Agile- an approach, principle, and way of thinking producing the right products and services for the customer and collaboration within cross-functional teams with each team member having autonomy to do something better or create something new(19). SCRUM- a framework used to facilitate a project that is comprised of shorter sprints and smaller deliverables. The framework consists of rules, roles, and use of artifacts for the implementation of Agile projects. It employs an iterative approach in order to optimize predictability and control risk(19).

Chapter VIII | PREDICTIVE

Predictive is about leveraging data analytics to be better positioned to react to problems before they occur. This advances the transformation towards a critical thinking organization where staying out in front of problems and risks is engrained within the cultural DNA.

Being predictive is about being able to foresee the variables and thus alter the outcomes. There is an acronym used in the military referred to as METT-T which stands for Mission, Enemy, Time, Troops, Terrain. When it is used, it is used to mean that the outcome just depends on the variables. In the past, some of these variables were difficult to predict but today with Industry 4.0 technologies, the variables are becoming more and more easy to predict.

Caution Points for Being Predictive:
- Always use trusted and clean data.
- Don't lose focus on the end state for the Predictive Phase which is to reduce and prevent problems from occurring.
- Most all companies talk about creating organizational problem solvers. The flaw with this thinking is that solving problems indicates that the problem has already occurred which places the company in a reactive state to solve the problem. The Predictive Phase is focused on staying out in front of problems and the financial impact that occurs as a result. It is the difference between a leading (predictor) and lagging (has already occurred) indicator.

Targeting the opportunities for Predictive tool application comes as a result of understanding where the risk is. Great practices to use are using a pre-mortem strategy to think of worst case scenarios and working backwards to resolve or even conducting situational analysis as a means to be prepared with an emphasis on speed and focus. Access the risk and apply Predictive tools to mitigate or even eliminate the risk.

PREDICTIVE TOOLS
1. Reliability Centered Maintenance (RCM)
2. Simulation
3. Data Analytics

1. Reliability Centered Maintenance (RCM):

RCM is a component of asset management that has the objective of predicting equipment failure through a focus on how the equipment fails (failure mode), the effects of the failure (failure effect), and the solution to the failure as to prevent reoccurrence. Elements include:

- Computerized Maintenance Management System (CMMS)
- Scheduling / Planning
- Corrective Maintenance
- Preventive Maintenance
- Proactive Maintenance
- Predictive Maintenance
- Spare Parts Management
- Asset Replacement
- Metrics & Assessment / Maturity Model

This tool is categorized in the Predictive Phase as it is designed to improve asset availability and reliability through predictive strategies and techniques. Companies focused on risk-reduction will use this tool to ensure asset availability and operation reliability.

Implementing strategies to prevent equipment failures is a key component of any asset performance management initiative. Failure prevention can be successful, however, only if those strategies address the underlying cause of equipment failures. Then, effects can be addressed and recommendations can be made for ways to prevent potential failures and avoid the consequences in a consistent, repeatable manner. As failures occur, this is an indication that reliability is decreasing which increases the need for maintenance. The need for maintenance increases maintenance costs. Conversely, as reliability increases, the need for maintenance decreases which then decreases maintenance costs. It is common knowledge within Maintenance communities that 5-10% of efficiency is lost if assets are not being monitored.

Caution Points- Where Companies Can Get This Wrong:

- Using data that are not trusted and clean during the data collection phase of deploying RCM.
- Not adhering to Preventive Maintenance (PM) and Predictive Maintenance (PdM) schedules.
- Not having enough resources to maintain a RCM program

- Poor communication and relationship between Operations and Maintenance.

Note: Use Cases (Application Benefits) located in the Appendix

Tool Box:

• RCM A3	• FMEA (Failure Mode and Effects Analysis)	• CMMS (software)
• TPM (Total Productive Maintenance)	• PF-Curve (Potential Failure- Curve)	

2. Simulation:

A simulation is a model that imitates the operations and flow of an existing process or a proposed process. The objective is to view the process with a focus on identifying:
- Barriers to flow
- Monuments
- Distancing / Spacing

This tool is categorized in the Predictive Phase as companies that seek to identify impediments to process flow prior to finalizing the process, will utilize this tool.

Running simulations allows you to work out the potential barriers to flow within a process before deploying capital. Some companies will use plastic building blocks to train employees how to simulate processes to identify these areas for opportunity. Reference the 3P Tool in Chapter V: Simply Phase.

Caution Points- Where Companies Can Get This Wrong:

- Not spending enough time running simulations prior to dropping utilities and pushing forward with other factory startup activities such as mounting equipment to floors.
- Not knowing their processes to begin with.
- Not factoring in WiFi coverage and the impact of poor coverage. This will be covered in Chapter IX: An Approach Supported by Industry 4.0.

Note: Use Cases (Application Benefits) located in the Appendix

Tool Box:

• 3D Scale Model	• CAD (software)

Additional Tools:

- Machine Order of Operations Analysis- used to dissect how machines interact with product through the study of machine timing and motion. **(Figure 8-2)**

MACHINE ORDER OF OPERATIONS ANALYSIS EXAMPLE

New/Rev _____		Page 1 of 1		Date: 00/00/00	Machine number: 001			
Organization: XYZ	Dept.		Leader	MACHINE ORDER OF OPERATIONS ANALYSIS	Machine name: welder			
Prod. Line: A	ABC		RAC		Machine cycle time: 56 sec.			

Order #	Component	Motion	Time Analysis Motion	Product Contact	Control Auto	Control Manual	No. Of Operators
1	Cradle transverse cyl.	Cradle indexes to C side (with shield)	4	X	X		0
2	(4) centering cyl.	Shield is centered	2	X	X		0
3	OH transfer	Frame lowered into cradle on top of shield	3	X	X		0
4	(4) clamping cyl.	Shield/frame clamped	3	X	X		0
5	(4) Z-pos gun cyl.	Z-position guns extend, weld, retract	3		X		0
6	(1)A/(1)D pos. Gun cyl.	Z-position guns extend, weld, retract pt. 1	2		X		0
7	(1)A/(1)D pos. Gun cyl.	Z-position guns extend, weld, retract pt. 2	2		X		0
8	(1)A/(1)D pos. Gun cyl.	Z-position guns extend, weld, retract pt. 3	2		X		0
9	(1)A/(1)D pos. Gun cyl.	Z-position guns extend, weld, retract pt. 4	2		X		0
10	(4) clamping cyl.	Shield/frame clamped released	3		X		0
11	OH transfer	Shield/frame assembly raised out of cradle	4	X	X		0
12	Cradle transverse cyl.	Cradle indexes to B side (no shield)	4		X		0
13	Manual switch	Shield C/V placed into manual mode	1			X	1
14	Human	(1) stack of (8) shields loaded	5	X		X	1
15	Auto switch	Shield C/V placed into auto mode	1			X	1
16	C/V transverse cyl.	Shield C/V indexes to B side	5	X	X		0
17	Welder OH transfer	Shield removed upward	3	X	X		0
18	Welder OH transfer	Shield indexes to D side	4	X	X		0
19	Welder OH transfer	Shield lowered into cradle	3	X	X		0
			54	10	16	3	3

cyl. = cylinder | pt. = point | OH = overhead | C/V = conveyer

SECTION IV | SYNTHESIS

Chapter IX | AN APPROACH SUPPORTED BY INDUSTRY 4.0

As companies work to transform their business, there must be a systematic approach to the desired future state. This approach is The 4 Phase Approach Model and to leverage its max impact, it does require support by Industry 4.0 technologies and solutions. However, in order to implement Industry 4.0 successfully, the organization's culture must be prepared and committed and the processes must be in a state of control. The level of commitment and control are provided through the Model.

So what is Industry 4.0? Before I answer that, let me provide some detail on Industry 1.0 through 5.0:
- Industry 1.0 (~1780)- Mechanization, water, and steam power
- Industry 2.0 (~1870)- Mass production, assembly line processing, electrification
- Industry 3.0 (~1970)- Computers, electronics, automation
- Industry 4.0 (~2011)- Machine learning and connectivity, IoT, robotics, etc.
- Industry 5.0 (~2020)- Human-machine collaboration, customization, cognitive systems

Even though we are in the Industry 5.0 era, most companies have not yet leveraged Industry 4.0 as of 2023. It is critical for companies to have a deep understanding of 4.0's applications, advantages, disadvantages, and required capital investment prior to implementing. Until this occurs, caution should be taken engaging 5.0. For this reason, I will direct the focus to Industry 4.0.

Industry 4.0 is about connections...
- Factory information technology with operational technology
- Machine to machine
- Product lifecycle across the value stream

It provides organizational speed through a focus on data and data processing and analysis. How data is utilized and analyzed provides insights for growth. Industry 4.0 is comprised of the technologies shown in the figure:
(Figure 9-1)

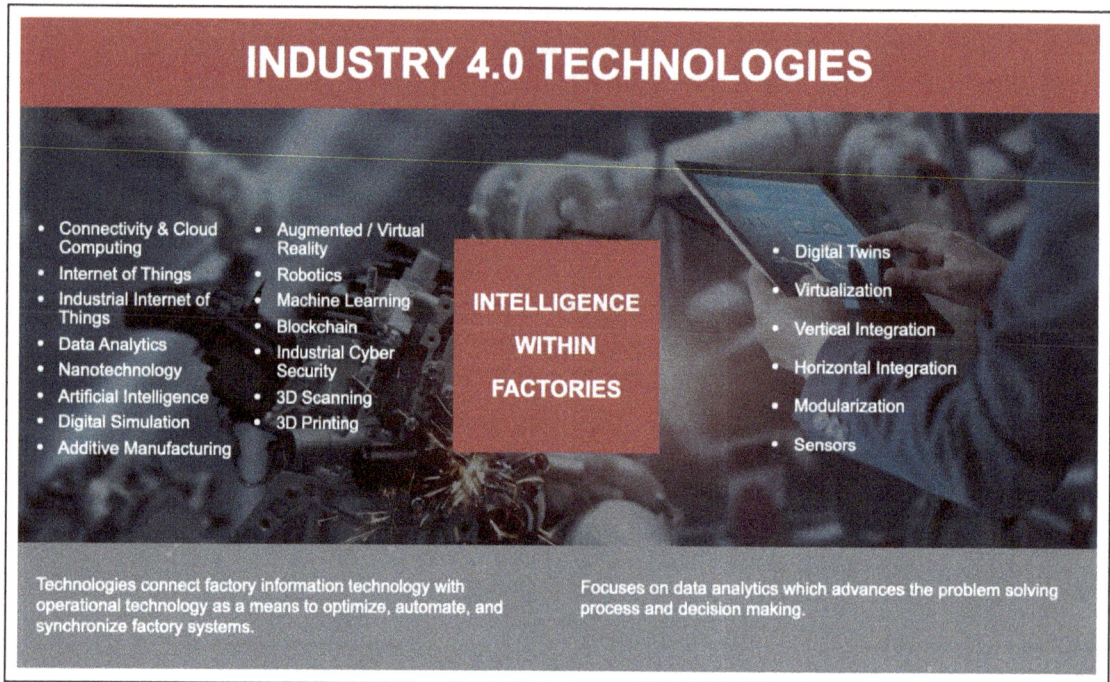

INDUSTRY 4.0 TECHNOLOGIES

- Connectivity & Cloud Computing
- Internet of Things
- Industrial Internet of Things
- Data Analytics
- Nanotechnology
- Artificial Intelligence
- Digital Simulation
- Additive Manufacturing

- Augmented / Virtual Reality
- Robotics
- Machine Learning
- Blockchain
- Industrial Cyber Security
- 3D Scanning
- 3D Printing

INTELLIGENCE WITHIN FACTORIES

- Digital Twins
- Virtualization
- Vertical Integration
- Horizontal Integration
- Modularization
- Sensors

Technologies connect factory information technology with operational technology as a means to optimize, automate, and synchronize factory systems.

Focuses on data analytics which advances the problem solving process and decision making.

The benefits of Industry 4.0 tools are as follows:
- Quicker time to problem solve / decision making
- Quicker than Industry 3.0 technologies
- Risk mitigation
- Enables innovation

There are some obstacles associated with the implementation of Industry 4.0 such as:
- Data availability
- Data quality
- Privacy concerns
- Willingness to invest
- Resistance to change

Some people may consider Industry 4.0 an obstacle to Lean but I feel Industry 4.0 takes Lean to higher levels. This isn't about pushing the boundaries of technology but rather leveraging existing technologies such as 3D scanning &

printing, machine learning solutions with RCM programs, AI language models as just a few examples. I also believe that companies that resist Industry 4.0 and now even 5.0 technologies will eventually be left behind.

The role that Industry 4.0 tools play in supporting The 4 Phase Approach Model is critical. **(Figure 9-2)**

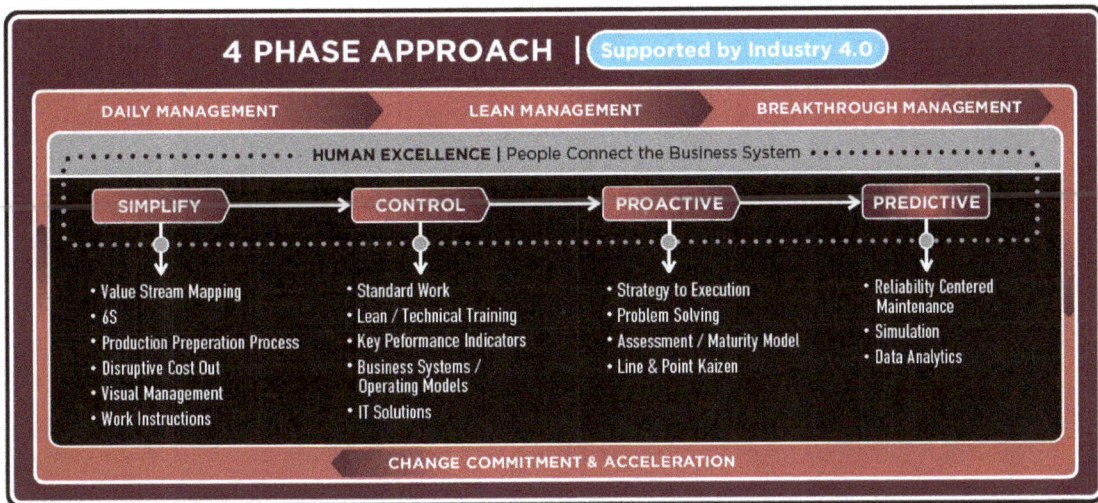

As a point of caution, I would create a multi-year plan to phase in Industry 4.0 technologies. As an example a company may start off with installing machine sensors to capture temperature, vibration, pressure, and rotation data then move to augmented and virtually reality for work instructions and training and by year 3 or so they are investing in robotics to eliminate repetitive work. There should be an identified problem to solve or a process to improve for the application of Industry 4.0 solutions.

Process steps for a company to begin to implement Industry 4.0 technologies such as AI are as follows:

1. Understand Business Goals and Challenges- Identify and understand what problem(s) the company is trying to correct and processes to improve. The 4 Phase Approach Model- Simplify Phase
2. Conduct a Current State Map and Assessment- Identify the as is state of the processes and evaluate where the opportunities for Industry 4.0 technology solutions exist. The 4 Phase Approach Model- Simplify Phase

3. Ensure the Processes are in a State of Control. Confirm processes are in control prior to applying Industry 4.0 technology solutions. The 4 Phase Approach Model- Simplify and Control Phases
4. Leverage Use Cases- Use AI applications (internal and external) for best practices and lessons learned. The 4 Phase Approach Model- Simplify Phase
5. Ensure Data Availability and Quality- Determine if adequate data exists, where it is located, and the cleanliness (quality) of it. Algorithms must be in place to manage the data. The 4 Phase Approach Model- Simplify Phase
6. Determine Budget and Resource Utilization- Determine the investment spend and match resources to ensure execution of the Detailed Implementation Plan. The 4 Phase Approach Model- Simplify Phase
7. Construct the Business Case- Determine the return on investment (ROI) factoring in any operational disruption. Determine where and when to apply. The 4 Phase Approach Model- Simplify Phase
8. Construct a Risk Assessment and Compliance Plan- Identify risks associated with Industry 4.0 technology implementation such as compliance and data confidentiality and security. The 4 Phase Approach Model- Simplify and Predictive Phases
9. Establish a Steering Committee and Core Team- Identify those who will provide guidance and direction and those that will lead the day-to-day implementation activities. The 4 Phase Approach Model- Simply Phase
10. Determine the Cultural State (Compliant, Committed, or a Hybrid) if not Already Known- Identify where potential resistance may reside and create a plan to mitigate / eliminate it. The 4 Phase Approach Model- Change Commitment & Acceleration
11. Create a Detailed Implementation Plan- Create the action roadmap to close the gap between the current state and desired future state. The 4 Phase Approach Model- Simplify Phase
12. Identify the Industry 4.0 Enablers- Match Industry 4.0 technology solutions to problem's root causes. Included within the Detailed Implementation Plan. The 4 Phase Approach Model- Industry 4.0
13. Communicate the Business Case- Present to the organization for input, transparency, and buy-in. The 4 Phase Approach Model- Change Commitment & Acceleration
14. Confirm Alignment- Ensure organizational buy-in prior to executing the Detailed Implementation Plan. The 4 Phase Approach Model- Human Excellence, Proactive, and Predictive Phases

15. Execute the Detailed Implementation Plan- Complete the identified actions to achieve the future state utilizing Industry 4.0 technology solutions. The 4 Phase Approach Model- Proactive Phase
16. Pilot Industry 4.0 Technologies- Confirm technology solutions through simulations. The 4 Phase Approach Model- Simplify and Predictive Phases

I am confident that if the recommended process steps to implement Industry 4.0 technologies are followed, successful results will be achieved.

Chapter X | PULLING IT ALL TOGETHER

Up to this point in the book, we have walked through each Phase independent of the others. Each Phase provides value but now let's pull them together where we multiply the value and thus maximize competitive advantage for the company.

The 4 Phase Approach Model must begin with a simplified approach and end with a predictable outcome. This provides an ongoing transformation leading to organizational growth.

The 4 Phase Approach Model is a proven model that does not work without the Human Excellence component of it; People Connect the Business System and Change Commitment & Acceleration. In addition, the Model must be supported by Industry 4.0 technologies. **(Figure 10-1)**

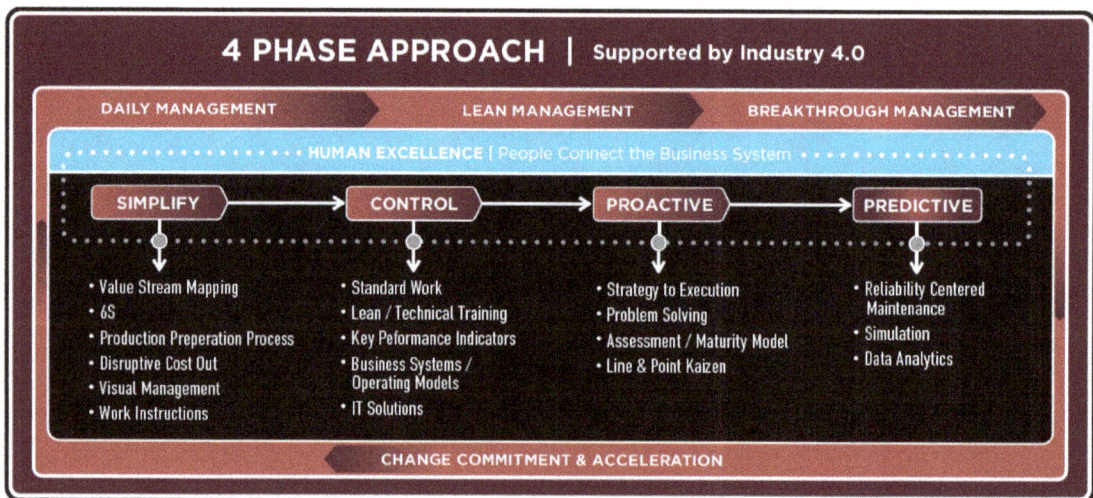

4 PHASE APPROACH | Supported by Industry 4.0

DAILY MANAGEMENT | LEAN MANAGEMENT | BREAKTHROUGH MANAGEMENT

HUMAN EXCELLENCE | People Connect the Business System

SIMPLIFY → CONTROL → PROACTIVE → PREDICTIVE

SIMPLIFY	CONTROL	PROACTIVE	PREDICTIVE
• Value Stream Mapping	• Standard Work	• Strategy to Execution	• Reliability Centered Maintenance
• 6S	• Lean / Technical Training	• Problem Solving	• Simulation
• Production Preperation Process	• Key Peformance Indicators	• Assessment / Maturity Model	• Data Analytics
• Disruptive Cost Out	• Business Systems / Operating Models	• Line & Point Kaizen	
• Visual Management	• IT Solutions		
• Work Instructions			

CHANGE COMMITMENT & ACCELERATION

APPENDIX

1. Value Stream Mapping (VSM)
Use Case #1 (Application Benefits)
New Product Introduction (NPI)

Business Case:
The CEO set a target to improve productivity across Engineering and program phase gates. The target was to reduce cycle time by 20% and improve First Pass Yield (FPY) by greater than 10%. The event then would become a model for future program projects.

Process Applied:
- Met with senior leaders to discuss baseline program to use / requirements for the VSM event.
- Identified the event team representing Commercial, Program Management, Engineering, Supply-Chain, Operations, and Quality.
- Conducted pre-work in lieu of the VSM event which included current state data and lessons learned based on best of the best (BoB) and the worst of the worst (WoW) programs.
- Day #1 of the event - kickoff with executive comments, introduction to Lean and Value Stream Mapping. Using different color sticky notes to represent process steps, deliverables, and improvements, the current state process began with working from the customer backwards asking "what step comes next?" Captured parallel steps and processes.
- Day #2 of the event - completed the process flow and the information flow. The time flow began by breaking the participants into three teams and assigning each team to one section of the current state map. Each team was tasked with calculating the cycle time (calculated from the end of process 1 to the end of process 2 and measured in days), applied time (measured in hours), inventory (measured in number of programs), and yield (measured in First Pass Yield or FPY) for their respective sections. The data were then entered into a data box per process step.
- Day #3 of the event - used red yarn to signal rework loops. From this point, the participants ran a simulation to ensure nothing was missed, overlooked, or misrepresented and most importantly, represented what they actually experience during each program.

- Day #4 of the event - identified current state key improvements and initiated the future state map. The future state map was created, factoring in opportunities identified from the current state map, which included areas for waste and risk reduction and opportunities to implement innovation.
- Day #5 of the event - ranked the improvement opportunities by using a Impact & Ease of Implementation Matrix. **(Figure 5-4)**

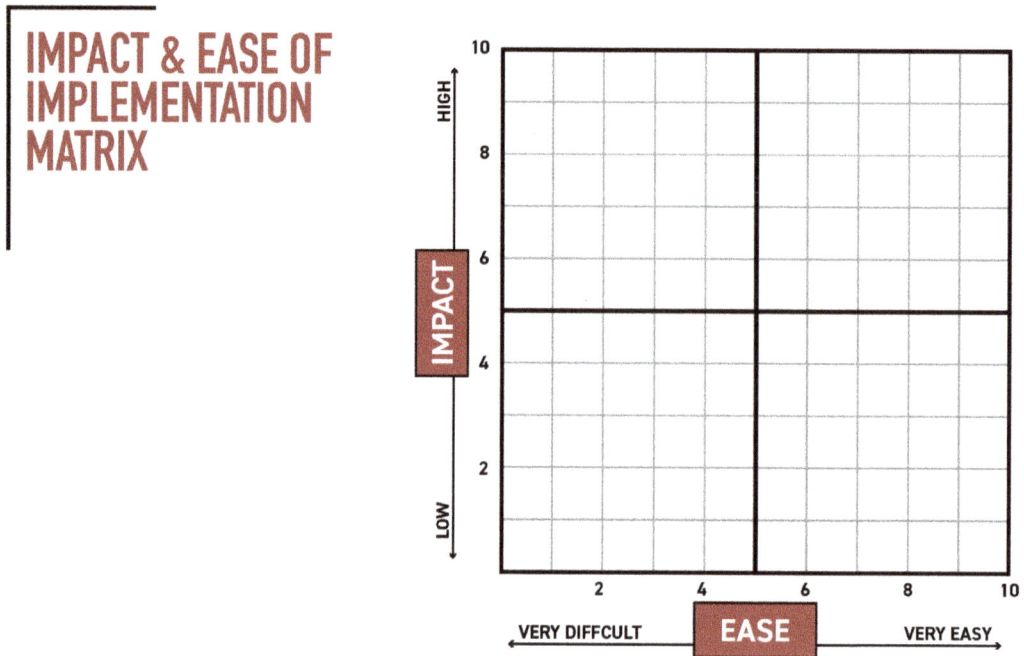

IMPACT & EASE OF IMPLEMENTATION MATRIX

The rule of thumb that was used was as follows:
Low Value / Low Effort- Debate
High Value / Low Effort- Just Do It (JDI)
High Value / High Effort- Create Project A3s
Low Value / High Effort- Kill it

The participants voted on the top four opportunities. All remaining opportunities were placed on a Parking Lot List for later use. **(Figure 5-5)**

PARKING LOT LIST

	PARKING LOT LIST					
Item #	Date	Event	Opportunity	Category: Daily | Lean | Breakthrough Mgmt.	Projected Time to Implement	Point of Contact

Once completed, four teams were created assigned to each of the four opportunities. Each team created a 4-Blocker slide comprised of:
- Problem Statement
- Scope & Measurement
- Project Objective / Goal
- Implementation Plan - included all actions required to complete the opportunity they were assigned to.

Once the 4-Blocker slide was completed, the event facilitator reported out to the executive staff. Each of the four teams met every two weeks post VSM event where they reported out on their progress towards the completion of all actions using action trackers.

Ah Ha Moments:
On Day #3 of the event once the current state map rework loops were identified, the commercial participants suddenly realized how bad of a situation they placed Engineering in. The process had illuminated that they were attempting to sell products that Engineering hadn't even figured out how to design yet. In addition, it became clear that customers were being permitted to make design changes, even after the manufacturing freeze date. The Engineering team were suddenly realizing how bad of a situation they were placing Manufacturing in

when they were outsourcing complex product drawings resulting in compatibility issues on the factory floor. Manufacturing were understandably irritated that they were positioned to fail with respect to customer on-time delivery.

Lessons Learned:
VSM Facilitation
Several leaders were not participating or interacting during the first two days of current state mapping. Once the current state map was completed and the rework loops identified, they began to take notice and witness where the issues were occurring. Note: this is the point I made earlier with respect to simplify the process by making it visible. In absence of doing this, you cannot see the issues and in this case, the organizational silos lead to misunderstanding.

Event Outcome
Functions are operating in silos and seldom communicating with one another.

Functions annual evaluation Key Performance Indicators (KPIs) are not aligned with one another. As an example, the Commercial Team may be evaluated on how many units they sell even if they are selling products that Engineering hasn't even figured out how to design yet.

End Result:
A 30% reduction in cycle time which exceeded the target set by 10% and improved FPY by 11%. This outcome would not have occurred if the value stream was not made visible.

Use Case #2 (Application Benefits)
Lab Life Testing Practices

Business Case:
Identify differences in lab life testing practices across three regions. Create a single standard process and counteract known issues relating to life testing practices that affect the number of steps, queue time, touch time, and rolled throughput yield (RTY).

Target: reduce the number of steps by >25% and RTY by >10%.

Process Applied:
- Met with the internal customer.
- Initiated pre-work.
- Conducted a pre-work meeting (4 weeks prior to the VSM event).
- Set up for VSM event Day #1.
- Day #1 of the event:
  - Presented opening presentation to the participants and leadership.
  - Team introductions.
  - Reviewed the VSM agenda.
  - Provided VSM and 8 Wastes Training.
  - Reviewed lessons learned.
  - Reviewed BoB and WoW for the current state.
  - Initiated the current state by asking "what is the first thing that happened?" Created the process and information flows.
  - Documented the current state lead time ladder and FPY data in a spreadsheet, which provided RTY.

- **Concluded Day #1** by asking the participants to add at least one feedback / comment per side of the Positives & Opportunities List wall display with respect to improving the event process **(Figure 5-6)**

Positives & Opportunities List				
Date	Positives +		Opportunities -	

- **Day #2 of the event:**
  - Opening remarks.
  - Reflected on Day #1 to ensure scope is still valid.
  - Reviewed Day #1 Positives & Opportunities.
  - Provided Waste Type Training.
  - Classified Day #1 identified issues and concerns into waste categories. These were from rework lessons learned and current state sticky notes.
  - Used Impact & Ease of Implementation Matrix to identify the wastes to be counteracted.
  - Divided high impact / low ease of difficulty for root cause identification
  - Formed sub teams.
  - Began developing counteractions for prioritized wastes using a Counteraction Template.
  - Assigned single point accountability (SPA) to each waste for corrective action execution post VSM event.
  - Sub teams presented to the group.
  - Created presentations for leadership detailing the week's progress up to the current state map completion as a means of seeking their buy-in. This was accomplished by simply asking them "Does the current state represent what you feel?"
  - Conclusion of Day #2 by asking the participants to add at least one feedback / comment per side of the Positives & Opportunities List
  - Took team picture.

- **Day #3 of the event:**
  - Opening remarks.
  - Reflected on Day #2 and mid-week executive report out (at this point, leadership will provide input for approval of current state VSM).
  - Reviewed Day #2 Positives & Opportunities.
  - Provided VSM future state training.
  - Initiated future state map.
  - Initiated executive report out presentation for Day #4.
  - Participants documented Day #3 Positives & Improvement Opportunities.

- **Day #4 of the event:**
  - Opening remarks.
  - Completed final executive report out prep and review with the participants.
  - Presented executive report out to the leadership team.

- **Post VSM event:**
  - VSM facilitator established a bi-weekly meeting cadence with SPA's for top identified wastes.
  - VSM facilitator established a monthly meeting cadence with leadership for SPA to report the executive plan.
  - VSM facilitator codified the VSM created in the event by entering it into an appropriate tool.

Ah Ha Moments:
- The amount of work associated with the rework loops was greater than the amount of work not associated with rework loops.
- Video recording the process, where applicable, and reviewing prior to initiating the current state VSM was very helpful.

Lessons Learned:

VSM Facilitation:
Working from a VSM standard operating procedure (SOP) was key to a successful event.

Event Outcome:
Regional labs were not communicating as "one" company lab. This led to an absence of best practice sharing, technical knowledge sharing, and discussions focused on problems, root causes, and implemented solutions.

End Result:
All regions working from a standardized process. The number of steps reduced by 40%, queue time reduced by 61%, touch time reduced by 72%, and RTY improved by 17%. All of these outcomes exceeded the projected goals.

2. 6S
Use Case #1 (Application Benefits)
Creating and Implementing 6S for Chemical Storage

Business Case:
The maintenance shop was not servicing and repairing trucks quick enough resulting in a backlog of repairs. The maintenance shop was determined that one of the causes was time taken by the technicians looking for chemicals required for servicing and repairs. The objective was to eliminate this one cause. Target: increase shop efficiency by 5%.

Process Applied:

1. Sort - After emptying the 5-shelf cabinet of all its chemical content, I sorted the items into categories such as paint, greases, penetrating oil, engine oil, transmission oil, antifreeze. I then determined what would not be placed back into the cabinet based on its expiration date, the ability to be used (for example, items with damaged nozzles were discarded), and the amount of chemical remaining in the container (consolidated where applicable). All chemicals that were not placed back into the cabinet were then red tagged and properly disposed of.
2. Set In Order - I determined how to organize the chemicals by chemical categories. For example, in my categorization, the greases were placed on the first shelf. The strategy was to place the most frequently used chemicals on the top shelf with the least used chemicals on the bottom shelf.
3. Shine - I cleaned the inside of the cabinet, followed by cleaning each of the chemical cans, bottles, and tubes that were remaining after the Sort Phase. The chemicals were then placed back into the cabinet.
4. Standardize - Next I created and attached a label on each shelf face which detailed what the chemical category and the minimum / maximum quantity were. This ensured that when the items reached the minimum quantity, we would order the necessary replenishments.
5. Sustain - I took a picture of the organized, cleaned, and labeled cabinet. I laminated the picture and attached it inside the cabinet door so that if it became unorganized again, the technicians would immediately know what it should look like.
6. Safety - The cabinet was placed on top of a metal bench that prevented the technicians from having to bend over or squat down to get to the chemicals on the lower shelves. In addition, I placed a material safety data sheets (MSDS) binder containing only the material safety data sheets for the chemicals in the cabinet on the side of the cabinet for easy reference.

Ah Ha Moment:

After using the cleaned and improved chemical storage unit, we discovered the opportunity to slant the shelves and install dividers that would make the minimum / maximum levels more apparent and the supplies easier to remove and replenish.

Lessons Learned:

There were new minimum / maximum levels established, thus preventing the previously frequent chemical overstocking and shortages.

End Result:
Target achieved: 5% efficiency improvement

Use Case #2 (Application Benefits)
Creating and Implementing 6S within Maintenance

Business Case:
Establish a 6S program for 20 Maintenance shops within 2 countries. The objective was to prevent future injuries due to poor housekeeping practices. Target: 0 recordable injuries.

Process Applied:
When I provided 6S training for maintenance technicians within the two countries, my message was not what you think it would be. I was a maintenance technician early in my career and I always viewed 6S as a practice that would keep me going home to my family in the same condition as when I left them; 6S was something that I took seriously. I had witnessed injuries due to poor housekeeping practices and I never wanted that to be me. When I conducted the training, I opened by stating that this 6S Program wasn't another example of management nit picking over trivial things. It was about being smart and looking out for their fellow technicians by the way they maintained the shops. I recounted a few of the harrowing injuries I witnessed as a result of not using 6S in the maintenance shops. I am sure they sensed my emotions coming through. I vividly recall seeing a safety video back in the late 90s when I was functioning as a process technician within the electronics industry called *Remember Charlie(12)*. This story was based on true events that changed the life of Charlie Morecraft. I still think about that story and remain vigilant about safety. Don't get me wrong, I was very safety conscious at that point. I spent 11 years as an aircraft maintenance crew chief on F-16, F-117, and KC-135 aircraft while serving in the United States Air Force. I was the guy on the headset talking to the pilot while the aircraft was running conducting flight checks prior to aircraft launch. That video took my safety consciousness to the next level. Once I reviewed what the 6S's were and established how important good housekeeping practices were and that I did not want to see anyone of them getting hurt, they got it and they walked away on a mission.

I walked through each maintenance shop within both countries once per month and provided my improvement input. I was amazed as to the individual shop 6S ownership and innovative thinking. The teams really cared about how their respective shops not only looked but how safe they were.

During my shop 6S tours, I would take pictures of 6S innovative thinking actions, insert them into a 6S Innovative Picture Template **(Figure 5-7)** and share them with all the shops across the company as a means to leverage the improvement ideas and best practices.

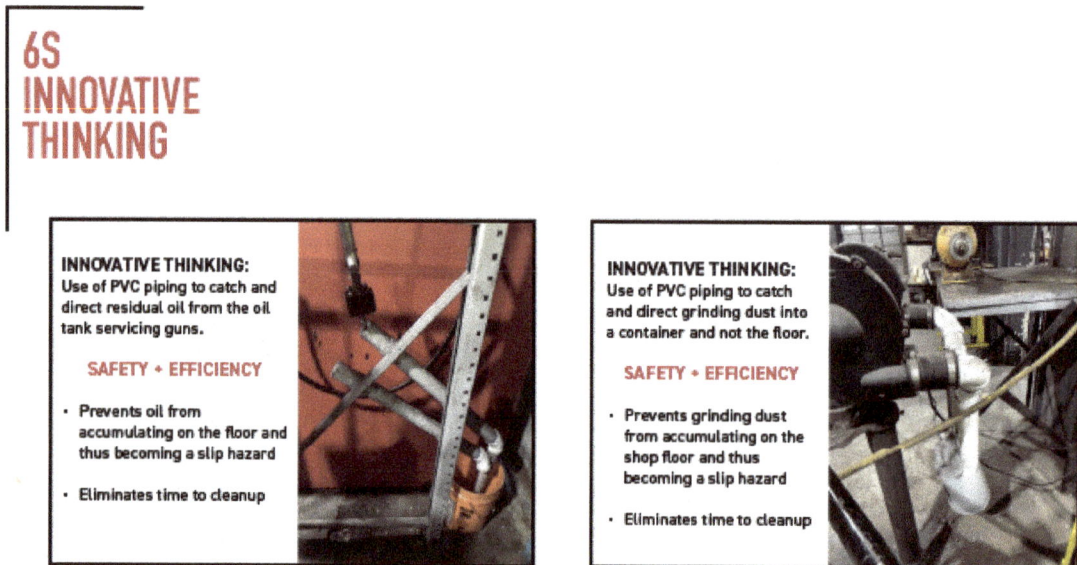

6S INNOVATIVE THINKING

INNOVATIVE THINKING:
Use of PVC piping to catch and direct residual oil from the oil tank servicing guns.

SAFETY + EFFICIENCY

- Prevents oil from accumulating on the floor and thus becoming a slip hazard

- Eliminates time to cleanup

INNOVATIVE THINKING:
Use of PVC piping to catch and direct grinding dust into a container and not the floor.

SAFETY + EFFICIENCY

- Prevents grinding dust from accumulating on the shop floor and thus becoming a slip hazard

- Eliminates time to cleanup

Ah Ha Moment:
The technicians generated high quality, innovative ideas to improve their use of the machines and tooling.

Lessons Learned:
Forcing 6S creates compliance. Working with each shop and providing them the business case, support, training, tools, and time they need to be successful creates commitment.

End Result:
Within 3 months of the initial training, every shop had completed the first 3S's: Sort, Set in Order, and Shine. Ownership of the program across the shops was evident. As a result, there were no injuries that occurred within the shops due to poor housekeeping practices during that time period

3. Production Preparation Process (3P)

Use Case #1 (Application Benefits)
New Factory Brownfield Design / Layout Using 3P

Business Case:
Install a state-of-the-art / intelligent Lean remanufacturing facility at a brownfield site using 3P. This facility will reduce costs associated with outsourcing remanufacturing activities. Target: startup within four months not to exceed a budget of $45 million dollars.

Process Applied:
Applied the Shingijitsu Production Preparation Process (3P) as detailed under this tool section.

Ah Ha Moment:
Running the simulation for the first time using the large scale mockup.

Lessons Learned:
- Stay focused on the event scope.
- Calculate the quantity of operators per station prior to laying out the process.
- Run simulations to ensure no process steps are missed. Always re-run simulations if you are deviating from the original 3P design such as replacing a horizontal machine with a vertical machine. Simulations ensure you are purchasing the correct equipment the first time.
- No question is a dumb question.
- There is power in try-storming (brainstorming with taking action such as prototyping).

End Result:
Achieved the time target and came in under the $45M dollar budget.

Use Case #2 (Application Benefits)
New Factory Greenfield Design / Layout Using 3P

Business Case:
Create a new manufacturing factory (greenfield site) in order to increase capacity for forecasted volumes. Target: startup within one year not to exceed a budget of $200 million dollars.

Process Applied:
- Applied the Shingijutsu Production Preparation Process (3P) as detailed under this tool section.

Ah Ha Moment:
Conducting a 3P off site can be accomplished as long as the landscape footprint detail is known.

Lessons Learned:
- Ensure those that are in charge of contracting the placement of machines, utility drops, and designated areas for gantry cranes (normally either the Facilities Manager, Engineering, or Maintenance) are in the communication loop with the remaining members of the 3P Team with respect to the simulated and approved layout. If this does not occur, expenses will be incurred to reverse what has been authorized or leadership will make the call to revise the 3P layout working around "out of process" authorizations.
- The due diligence up front during the 7 Ways is critical for the ideal end result. The rules of 7 Ways must be followed in order to deliver the best outcome.

End Result:
Achieved the time target and came in under the $200 million dollar budget.

4. Disruptive Cost Out (DCO)
Use Case #1 (Application Benefits)
Gain Market Share through Disruptive Cost Out

Business Case:
Identify how to remove cost per manufactured unit in order to gain market share. Target: identify $250,000 of cost per unit to be eliminated.

Process Applied:
- Created a DCO Steering Committee.
- Created the DCO Business Case and Project Charter.
- Broke the product down into seven sections.
- Created teams per section comprised of: employees who work in the areas of the factory on the specific sections, suppliers / vendors, former employees, and employees who do not work in areas of the factory on the specific sections.

- Launched into DCO weekly meetings where each system, sub-system, and component fit / form / and function were placed under a microscope.
- Identified areas of opportunity that would reduce cost but not affect product function and safety.
- Quantified the financial cost savings.
- Presented to the executive staff using a 4-Blocker slide comprised of an Idea Description, Savings, Investments & Timing, Picture or Sketch of Idea, and Key Technical / Customer Challenges sections.

Ah Ha Moment:
Always review what you are paying your suppliers to do. In one case, we were paying for a bracket to be welded onto a section of chassis that we were no longer attaching anything to since the 80's. This represented ~$700 per unit of cost.

Lessons Learned:
Challenge every aspect of manufacturing a unit. If you are not manufacturing a spacecraft, your tolerances do not have to be as tight as required for space flight. Watch out for the "gold plating". This condition is categorized as over processing which is one of the 8 Wastes.

End Result:
Identified actions to remove $253,000 of costs per unit exceeding the target

Use Case #2 (Application Benefits)
Gain Market Share through Disruptive Cost Out

Business Case:
Identify how to remove cost per manufactured unit in order to gain market share. Target: identify $110,000 of cost per unit to be eliminated.

Process Applied:
- Created a DCO Steering Committee.
- Created the DCO Business Case and Project Charter.
- Broke the product down into five sections.
- Created teams per section comprised of: employees who work in the areas of the factory on the specific sections, suppliers / vendors, former employees, and employees who do not work in areas of the factory on the specific sections.

- Launched into DCO weekly meetings where each system, sub-system, and component fit / form / and function were placed under a microscope.
- Identified areas of opportunity that would reduce cost but not affect product function and safety.
- Quantified the financial cost savings.
- Presented to the executive staff using a 4-Blocker slide comprised of an Idea Description, Savings, Investments & Timing, Picture or Sketch of Idea, and Key Technical / Customer Challenges sections.

Ah Ha Moment:
Once the team began identifying items (cost) that were not required for the manufacturing of the product, it became exciting to see how much more cost could be pushed out.

Lessons Learned:
Do not assume that everyone on the team completely understands the mission of conducting the DCO. It is helpful to have a beginning of the day meeting around an Obeya Wall (Japanese word for war room or war wall) **(Figure 5-12)**

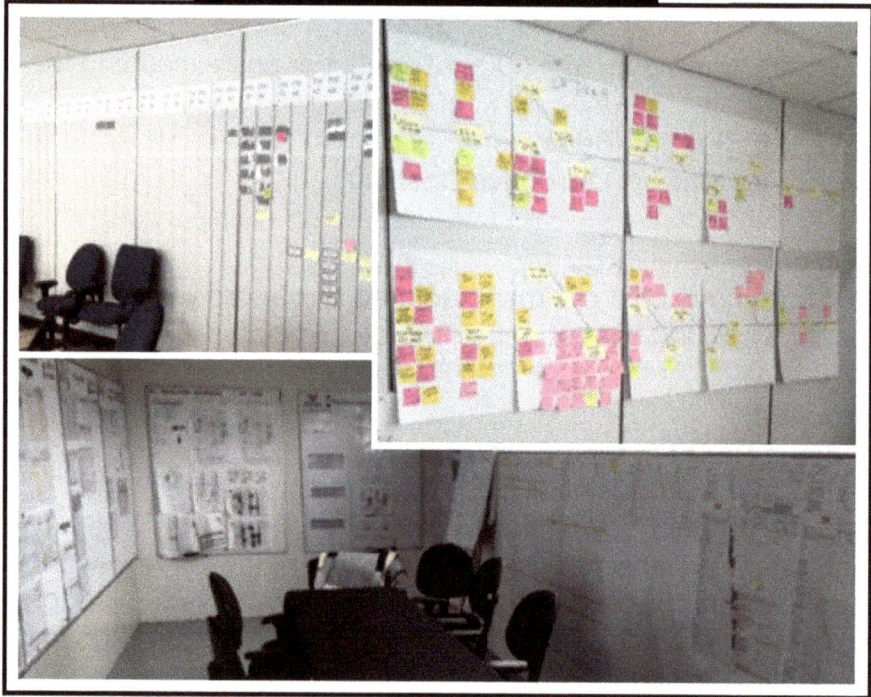

OBEYA ROOM EXAMPLES

where the objectives for the day are discussed and a meeting at the end of the day for each person to document on sticky notes what they felt went well and did not go well occurs.

End Result:
Identified actions to remove $116,000 of cost per unit exceeding the target

5. Visual Management
Use Case #1 (Application Benefits)
Work Instructions for Improved Efficiency

Business Case:
Increase process efficiency by eliminating the need to read work instructions (standard work) for newer employees to the process. Objective: Replace work instructions in the paper form with Augmented Reality (AR). Target: increase efficiency by between 5% and 10%.

Process Applied:
- Contracted with a company who specializes in AR.
- They video-recorded an employee working through a task that was performed correctly to the established work instruction.
- Once the recording was completed, the employee was replaced in the video with an avatar.
- The work instruction content was then inserted into the video via voice-over.
- Once the AR was completed for all the tasks within the specific station, the employee(s) would simply use a tablet to guide them through the tasks.

Ah Ha Moment:
Most of us when faced with a task within our personal lives such as working on a car, lawn equipment, or a home improvement project will grab our smart phone, tablet, or laptop and go online to locate a video of someone doing the task. The AR process applied the same logic.

Lessons Learned:
Feedback from employees validated that they enjoyed their work more having the AR capability to assist them with their jobs. In addition, it proved its worth when training new employees or cross-training employees.

End Result:
Efficiency gain of 8% was accomplished and employee turnover reduced due to work instructions in place and enhanced training.

Use Case #2 (Application Benefits)
Communication through Digital Daily Management Boards

Business Case:
Improve employee engagement by implementing digital Daily Management Boards (DMBs). Target: increase employee survey score with respect to communication as compared to the previous survey.

Process Applied:
- The company contracted with a vendor who provided digital DMBs. The digital boards replaced the boards that contained dry erase and paper updates. The digital platform enabled much more content to be viewed via touch screen.
- Training provided by the vendor to the process supervisors and other designated employees who would be required to enter updates.
- Supervisors conducted meetings with process employees to walk through the new boards to ensure everyone was aware of them and how to use them.
- The Director of Operations conducted monthly audits to ensure the DMBs were being updated as agreed upon.

Ah Ha Moment:
Employee reaction and supervisor engagement with the boards were positive. They felt that the boards were telling the story of the process current state. One employee stated it was like the shop floor was talking.

Lessons Learned:
- Be cautious not to add more work updating the digital boards than the manual updates.
- Visual Management without management support and standard work turns into nothing more than wall art.

End Result:
Feedback from the employees was that the digital visuals generated more employee engagement. There was a year-over-year employee survey score improvement with respect to communication.

6. Work Instructions

Use Case #1 (Application Benefits)

Create Work Instructions as Part of a Factory Startup

Business Case:

Create and implement work instructions per process as part of the 3P Event. Target: complete the creation of work instructions, provide training, and implement prior to scheduled startup. Note: This is categorized as foundational as it becomes an enabler for other actions to drive financial impact.

Process Applied:

- Created standard work.
- Used the standard work to create detailed work instructions per process. We used video and pictures taken during the 3P Event.
- Once created, leveraged Industry 4.0 AR technology to merge video of each task being completed correctly with the content of the work instructions. The employee completing the task was replaced with an avatar and the work instruction content was integrated into the video through voice-over.
- Conducted feedback sessions with the employees using the AR technology to ensure buy-in.

Ah Ha Moments:

Consider outsourcing the creation of the work instructions if resources are not available.

Lessons Learned:

- Creating work instructions is time consuming and requires a lot of effort. Ensure this work is being monitored in order to ensure completion and availability of the work instructions as per the established factory startup schedule.
- Once work instructions have been created and in initial use, there are always small updates that need to be made. The instructions should be treated as living documents.

End Result:

All factory work instructions created, validated, and in place for start of production.

Use Case #2 (Application Benefits)
Create Work Instructions within an Existing Factory where None Currently Exists

Business Case:
Create and implement work instructions per process. Target: complete the creation of work instructions, provide training, and implement prior to scheduled startup. Note: This is categorized as foundational as it becomes an enabler for other actions to drive financial impact.

Process Applied:
- Ensured the standard work was created, up-to-date, and implemented.
- Used the standard work to create detailed work instructions per process. We used video and pictures taken during the 3P Event.
- Once created, leveraged Industry 4.0 AR technology to merge video of each task being completed correctly with the content of the work instructions. The employee completing the task was replaced with an avatar and the work instruction content was integrated into the video through voice-over.
- Conducted feedback sessions with the employees using the AR technology to ensure buy-in.

Ah Ha Moments:
Leveraging AR

Lessons Learned:
- Maintain a file with each process requiring work instructions and report out on a weekly basis through Red / Yellow / Green (RYG) status. This ensures that the work instructions are being created, reviewed, and validated according to a schedule.
- Ensure that everyone involved has received training on standard work and work instructions prior to creating the work instructions.

End Result:
All factory work instructions created, validated, and in place for start of production. Employee retention improved as a result of standardized work instructions and use of AR.

1. Standard Work

Use Case #1 (Application Benefits)

Create Standard Work For a Remanufacturing Factory Startup

Business Case:
Create standard work across the factory for remanufacturing of a product. Target: create, train, and implement required standard work prior to scheduled factory startup. Note: This is categorized as foundational as it becomes an enabler for other actions to drive financial impact.

Process Applied:
- Completed the Production Preparation Process (3P) event.
- Using the following tools, we completed the detailed data gathering and analysis:
  - Time Observation Sheet
  - Standard Production Capacity Sheet
  - Standard Work Combination Sheet
  - Standard Work Sheet
  - Yamazumi Chart
- The new employees who were already hired and onboarded were included in the detailed data gathering and analysis.
- Once the body of work was completed, work instructions were created for each process. Work instructions were covered in Chapter V: Simplify.

Ah Ha Moments:
Using the product manufacturing standard work in essentially the reverse order for product remanufacturing.

Lessons Learned:
- If new hires cannot or are having difficulty understanding or following created standard work, investigate to better understand why and make adjustments, where applicable.

End Result:
All factory standard work created, validated, and in place for the scheduled start of production.

Use Case #2 (Application Benefits)
Create Standard Work Where None Exists Within a Manufacturing Factory

Business Case:
Create and implement standard work within a factory where no standards exist outside of work sequence steps and standard time per step. Target: create, train, and implement the required standard work across all factory processes. Note: This is categorized as foundational as it becomes an enabler for other actions to drive financial impact.

Process Applied:
- Completed a process Production Preparation Process (3P) event.
- Using the following tools, we completed the detailed data gathering and analysis:
  - Time Observation Sheet
  - Standard Production Capacity Sheet
  - Standard Work Combination Sheet
  - Standard Work Sheet
  - Yamazumi Chart
- The new employees who were already hired and onboarded were included in the detailed data gathering and analysis.
- Once the body of work was completed, work instructions were created for each process. Work instructions were covered in Chapter III: Simplify.

Ah Ha Moments:
The employees, to include leadership, did not understand what standard work was.

Lessons Learned:
New employees without standard work equates to employee retention issues.

End Result:
Selected one process for the standard work pilot then leveraged the body of work across the factory. Worked to establish the standards which were used for onboarding of many new employees.

2. Lean / Technical Training
Use Case #1 (Application Benefits)
Establish a Lean Training Program Within a Mature Company

Business Case:
Establish a Lean Training Program where known currently exists. Target: create and implement a Lean Training Program within a six month period. Establish an annual target for employees trained and certified. Note: This is categorized as foundational as it becomes an enabler for other actions to drive financial impact.

Process Applied:
- Identified training needs. In this case it was Lean Green Belt training.
- Selected an external training provider to provide the training.
- Communicated the training program with the executive leadership team then to the business unit leadership.
- Set a target per year of how many employees would complete the training and certification project.
- Launched the training.
- Communicated with the organization, through internal communication channels, the employees who successfully completed the training and project. The training and certification data were tracked by the external training provider.

Ah Ha Moments:
Behind closed doors, some business unit leaders were not promoting the training but rather deterring potential training candidates from committing to registering for the training.

Lessons Learned:
- Don't make training mandatory because it drives the wrong behavior. The business unit leaders know what they are accountable for so either they will leverage Lean to accomplish it or they won't. The ones who leverage it will identify the employees within their business they would like to complete the training. Those who don't rarely are successful.
- Virtual training was not ideal.
- The "one and done" mindset was present once the certification project was completed.

End Result:
Even though the targeted quantity trained and certified was achieved, the mandatory Lean Green Belt training drove the wrong behavior as it came across as forced and about making a quota.

Use Case #2 (Application Benefits)
Transition Training In-House

Business Case:
Transition Lean Green Belt Training from an external source to in-house. Target: complete the transition within a 3 month time period with a scheduled pilot launch. Note: This is categorized as foundational as it becomes an enabler for other actions to drive financial impact.

Process Applied:
- Communicated the transitioning of the training program in-house with the executive leadership team then to the business units leadership.
- Used the external source's training material and agenda.
- Conducted several meetings with the external source to ensure all components of the training program were covered.
- Created an internal training guidelines of all the requirements that must be met in order to begin the training. Captured lessons learned from the external training was used.
- Included more company specific examples within the training material.
- Ensured adequate pilot launch support from IT and Operational Excellence Administration.
- Launched the pilot training.
- Audited the pilot post training.
- Made all necessary audit finding changes.
- Provided additional support with student company projects.

Ah Ha Moments:
Students not completing a Project Charter prior to the initiation of the course. This was added to the training guideline list.

Lessons Learned:
- Don't make training mandatory… it drives the wrong behavior. The business unit leaders know what they are accountable for so either they will leverage Lean to accomplish it or they won't. The ones who leverage it will identify the employees within their business they would like to complete the training. Those who don't rarely are successful.
- Trainers need to understand the shortcomings of the tools.
- Teach individuals to effectively frame and scope problems, not to blindly implement solutions.

- Understand the root cause as to why a high percentage of students who completed the training but did not complete the project.
- Always ensure the training guidelines are reviewed with the student prior to the course start.
- A Lean 101 primer course can be used to get team members to see practical application of tools… this builds momentum. Demystifying Lean is critical.

End Result:
The pilot launched successfully and to the targeted schedule.

3. Key Performance Indicators (KPI)
Use Case #1 (Application Benefits)
Create a New KPI

Business Case:
Create a KPI that will capture a product that is shipped in several pieces for reassembly (kits) to an end customer (common practice with exporting of large product) with the goal of on-time delivery and also completeness of each delivery. As an example, the shipping company can ship a kit containing only 80% of the pieces / parts on-time and claim a 100% on-time delivery but it doesn't capture the incompleteness of the kit. Target: create a KPI to measure shipped complete and on-time. Note: This is categorized as foundational as it becomes an enabler for other actions to drive financial impact.

Process Applied:
- From a customer-supplier relationship perspective, brought the export shipping leaders and the customer together to better understand if the shipping company and the customer are measuring success the same way; is there alignment to how we measure success.
- Once that was determined, both parties agreed there was a need to create a KPI that captured on-time delivery and shipped complete.
- Created the KPI and included it in the Customer-Supplier Scorecard.

Ah Ha Moments:
The on-time delivery KPI was not viewed the same between the shipping company and the customer. The shipping company viewed an incomplete shipment shipped on-time as 100% on-time delivery whereas the customer viewed on-time delivery as 100% if all parts were delivered on-time.

Lessons Learned:
From a customer-supplier relationship perspective, all KPIs should be agreed upon by both parties.

End Result:
Created the Shipped Complete and On-Time (SCOT) KPI as a result of the customer-supplier collaboration.

Use Case #2 (Application Benefits)
Daily Management Board (DMB) KPIs

Business Case:
As part of a new factory start-up, there is a need to implement use of DMBs within each process that contain key KPIs to be reviewed each day as part of the leadership genba walks. Target: identify, standardize, and implement KPIs per process DMB within a three month time period. Note: This is categorized as foundational as it becomes an enabler for other actions to drive financial impact.

Process Applied:
- Installed DMBs within each process where they would be visible.
- Set up the boards following the S (safety) Q (quality), O (output), D (delivery) C (cost) KPI framework. Note: some companies will use A (associates), C (customer), and M (morale) as well. **(Figure 6-7)**

Daily Management Board

- Met with each SPA for each section to agree upon the leading and lagging indicators that would be included within the KPI Trackers that would be placed on the board.

- Ensured that the selected KPIs were aligned with and driving the company strategic plan.
- Posted the agreed upon KPIs on the DMBs.
- Met with the teams within each of the processes to review and provide them training for leading and lagging indicators within each section. The objective was to ensure everyone understood the differences between leading and lagging indicators, were aligned to how we were measuring success, and understood how what they did in their role moved the needle on each of the KPIs.

Ah Ha Moments:
The employees who were compliant but not committed were in that state as they didn't quite understand the purpose of the KPIs even after the initial communication and training.

Lessons Learned:
- Don't let updating the KPIs become just a something to update on a checklist, but rather ensure there is engagement.
- Don't let "I don't have the data readily available, so we can't have that as a KPI" be an excuse.
- Ensure each process understands how its output can impact the proceeding process (internal customer).
- Use the daily genba reviews of the KPIs to question unfavorable results and initiate root cause analysis and celebrate the favorable results. It is essential to understand the mechanics of how the process wins and loses.

End Result:
The DMBs were implemented within a three month time period. It took some time to have a standard level of engagement across all twelve processes but once it occurred, a problem solving culture was created.

4. Business System / Operating Model
Use Case #1 (Application Benefits)
Create a Business System

Business Case:
Create a business system within a company that has just implemented an Operational Excellence function and training program. Target: create a company business system to detail how the company should be structured and poised to

achieve strategic plan targets. Target: complete within a six month time period. In addition, measure employee understanding and engagement during the next employee survey as it relates to the business system. Note: This is categorized as foundational as it enables other actions to drive financial impact.

Process Applied:
- Reviewed the 5-year strategic plan, company vision / mission / values statements with the executive leadership team. The objective was to refresh any of the content, where applicable, and ensure alignment.
- Created the architecture of a business system focused on the key elements of people / process / performance taking into account the 5-year strategic plan and vision / mission / value statements.
- Reviewed the architecture with the executive leadership team in order to add additional detail and enhance the framework into a final business system.
- Once finalized, worked with the Marketing and Corporate Affairs teams to create the business system graphic.
- Reviewed the business system with the board of directors for alignment and approval.
- Once approved, created an internal marketing and communication blitz with all company employees to ensure employee understanding, drive excitement, and gain commitment.

Ah Ha Moments:
Do your due diligence in not creating a complicated business system within a small organization.

Lessons Learned:
- It is sometimes easier to create a business system with a new Chief Executive Officer (CEO) in place as compared to creating one with a CEO who has been in place for some time leading the organization without a business system.
- A less prescriptive business system is more adaptable to more situations.
- Top-down support must be coupled with great ideas generated from bottoms up.
- You cannot never over-communicate the importance of each employee understanding how what they do each day moves the needle within the business system.

End Result:

Business system in place within the targeted 6-month time period. Communication and audit plan in place to ensure ongoing understanding, alignment, and commitment to it.

Use Case #2 (Application Benefits)

Execution of a Business System

Business Case:

Drive the organization's in place business system. Target: increase contribution margin by >10% points within the next three years of the strategic plan. Note: This is categorized as foundational as it becomes allows other actions to drive financial impact.

Process Applied:

- Messaged the importance of and how the Business System drives value during every event where there were larger populations of employees in attendance such as Town Halls.
- Included the Business System graphic in the introduction of every training module.
- Conducted one-on-one messaging of the Business System with business unit leaders and functional leaders.
- Included employees opinion of the effectiveness of the Business System within the annual employee surveys. Some of the options to select from were:
  - Are you aware the company has a Business System?
  - Do you understand how the Business System works?
  - Do you understand the role you play within the Business System?
  - Do you see any value in the Business System?

Ah Ha Moments:

Not many employees within the organization, to include members of the executive leadership team, really understood the purpose or mechanics of the Business System.

Lessons Learned:

- Never assume everyone within the company understand the role they play within the Business System.
- There is some correlation between how innovative, "cool" and how the business system is marketed to the interest employees will have in it.

End Result:
The Business System never became "sticky" (accepted) enough to drive any value.

USE CASES PHASE | PROACTIVE

1. Strategy to Execution (STE)
Use Case #1 (Application Benefits)
Implement STE within a Company with Operations Across Five Regions.

Business Case:
Successfully implement STE across all 5 regions through training, workshops, mentoring, and facilitating monthly and quarterly reviews in addition to annual refreshes. Target: double revenue and segment income within the 5-year period in addition to expanding market share within two of the regions.

Process Applied:
- Followed the STE PDCA cycle (created the X-Matrix).
- STE initiative review process followed:
  - Initial STE review meeting
  - Reviewed the STE Dashboard

- Reviewed the Annual Strategic Initiative Action Register **(Figure 7-11)** from the previous STE Meeting

ANNUAL STRATEGIC INITIATIVE ACTION REGISTER			
ANNUAL STRATEGIC INITIATIVE	INITIATIVE OWNER	STATUS (R/Y/G)	NOTES

- Deep dived STE initiatives (scheduled and "red" status).
- When status was "red", reviewed Action Plan and KPI Tracker. Discussed: 1) actual vs. planned target, resources, risks, KPI results and initiative value, 2) issue root causes, 3) countermeasures, 4). corrective action plan, and 5) timing to recover and be back on track
- When status was not "red", reviewed Action Plan and KPI Tracker. Discussed: 1) actual vs. planned milestones and deliverables, 2) resource sufficiency to execute plan, 3) initiative risks, 4) actual vs. planned target KPI results, 5) current KPI trends, and 6) risks to planned value.
- Updated Action Register and Lessons Learned.

Ah Ha Moments:
STE 5-year executed goals will only achieve ~70% of the true north. The remaining ~30% will come from the organization's Daily and Lean Management.

Lessons Learned:
- It is difficult to get hundreds of people, if not thousands, to push in the same direction. As the facilitator, you must be capable of leading with and without authority and influence in order to be successful.
- The role of an experienced STE facilitator can contribute to a process' success.

End Result:
Generated $700M in revenue and $300M in segment income within the first 3 year period in addition to gaining market share within the two regions.

Use Case #2 (Application Benefits)
Drive Results within an Existing STE Structure

Business Case:
Drive an already established STE structure. Target: Increase contribution margin by 25 plus percentage points within the 5-year period.

Process Applied:
- Conducted Breakthrough Ideation sessions to identify annual ideas.
- Vetted and prioritized the ideas.
- Completed a business case analysis for ideas selected as potential annual initiatives that were used to drive the annual objectives.
- Determined if the potential initiative had a run rate impact and included within annual EBITDA budget and KPIs. If it did, it was categorized as a STE annual initiative. If it did not, it was categorized as a STE stretch initiative. Stretch initiatives were the "icing on the cake" initiatives as there was no established accountability for them but successful completion contributed to any fallout with the budgeted annual initiatives.
- Team took action to initiate and complete the initiatives during the course of the year.
- STE monthly reviews conducted to ensure favorable results. When favorable results were not achieved, corrective action was initiated.

Ah Ha Moments:
- If there is a newer CEO in place, ensure if the strategic plan is in need of a refresh under the new leadership.
- There are likely to be changes along the way.

Lessons Learned:
- If a company is small enough and are effective at working on annual initiatives with KPI tracking in place, there is no need to have additional X-matrices beyond the enterprise and business unit levels.
- By classifying an initiative as "run the business" as compared to "breakthrough", it does not mean it is unimportant. It just means it is managed differently.
- You have got to have a team committed to the true north vision and bringing their own connection in the execution of that vision.

End Result:
Increased contribution margin by 43 percentage points within a 4-year period.

2. Problem Solving
Use Case #1 (Application Benefits)
New Product Introduction (NPI)- Platform Development Cycle Time Reduction

Business Case:
Reduce New Product Introduction (NPI)- Platform Development Engineering cycle time. Target: 20% reduction at a savings of ~$3.4M

Process Applied:
- Establish a NPI Cycle Time steering committee and core team
- DMAIC Tools Applied.
    - Completed a Problem Definition Tree.
    - Completed a Project Transfer Function.
    - Conducted a 4 day VSM Event.
- Executive Kickoff.
- Current State.
- Executive Report Out.
- Future State.
- Detailed Implementation Plan.
- Executive Final Report Out.
- Standardized Control Plan.

- Prioritized the Detailed Implementation Plan by impact and effort and created four teams to execute the top four action items.
- Monthly action reviews were conducted to ensure the action items were being successfully executed.
- Change Commitment & Acceleration (CC&A) used as a means to gain project commitment through the CC&A Model phases:
  - Align
  - Envision
  - Buy-In
  - Sustain / Enhance
  - Measure

Ah Ha Moments:
Leaders from Commercial, Engineering, Supply-Chain, and Operations working within silos for years witnessing, through the Define and Measure stages, just how misaligned they were within one another to the end objective of delivering a product to a customer.

Lessons Learned:
- Don't let the quantity of tools overwhelm the team. Select a few from the tool box that will provide the desired outcome. An experienced Operational Excellence leader will be well-equipped to assist with the tool selection.
- Solutions are meaningless without knowing the question you are trying to answer.

End Result:
Engineering cycle time reduced by 30% at a savings of ~$5M

Use Case #2 (Application Benefits)
Material Availability (Make) Improvement

Business Case:
Improve the on-time delivery (OTD) of material make. Target: 10% improvement at a savings of ~$8M.

Process Applied:
- Establish a Material OTD steering committee and core team.
- DMAIC Tools Applied
  - Completed Team Charter / Primary Stakeholders

- Completed a Problem Definition Tree
- Completed a Project Transfer Function
- Conducted a four day VSM Event
  - Executive Kickoff
  - Current State
  - Executive Report Out
  - Future State
  - Detailed Implementation Plan
  - Executive Final Report Out
- Standardized Control Plan.
  - Prioritized the Detailed Implementation Plan by impact and effort and created four teams to execute the top four action items. Note: one of the action items was the creation of an inventory entitlement tool.
  - Monthly action reviews were conducted to ensure the action items were being successfully executed.
- Change Commitment & Acceleration used as a means to gain project commitment through the CC&A Model phases:
  - Align
  - Envision
  - Buy-In
  - Sustain / Enhance
  - Measure

Ah Ha Moments:
The Inventory Entitlement Tool was not a one size fits all tool across all the business units. The tool had to be customized to fit each business unit.

Lessons Learned:
- Ensure that the project team understands the metrics that are being used to measure progress.

End Result:
Material Make improvement of 14% at a savings of ~$14.1M

3. Assessment / Maturity Model
Use Case #1 (Application Benefits)
Improve Operations within a Pilot Business Unit

Business Case:
Improve Operations within the Pilot Business Unit. Target: Increase productivity by 3 to 5%.

Process Applied:
- Established a team comprised of Operations, Safety, and Quality leaders.
- Created the Operations vision and mission statements.
- Identified the Operations lagging and leading KPIs.
- Identified the Operations assessment pillars.
- Identified the Operations assessment elements per pillar.
- Determined the number of maturity levels which was 5.
- Created maturity behaviors per maturity level per element.
- Reviewed the model to ensure alignment.
- Determined the criteria weighting per pillar.
- Determined the total scoring guidelines.
- Using the newly created model, Operations conducted a self-assessment.
- Using the newly created model, Operations from another business unit conducted the peer assessment.
- Based on the scoring outcome, Operations selected three areas for opportunity to focus on during the course of the year with the target to mature the scoring for the following year's assessment.
- Areas for opportunity were included within the owners annual performance review.

Ah Ha Moments:
Link financial savings to each level of maturity, where applicable.

Lessons Learned:
- Always remember the "Why" you are doing this.
- Selecting the lowest rated 3 elements to focus on during the year is acceptable but note that Safety elements take priority.
- Pay attention to sequencing of the elements. As an example, you would want to ensure that Maintenance Strategy and Maintenance Planning were improved prior to improving Predictive Maintenance.

End Result:
Increased Operations productivity by 4.2%.

Use Case #2 (Application Benefits)
Improve Reliability and Availability

Business Case:
Improve reliability and availability. Target: Reduce maintenance costs by greater than $2.5M and improve equipment availability and reliability by 7 - 10%.

Process Applied:
- Established a team comprised of Maintenance and Engineering leaders.
- Created the Maintenance vision and mission statements.
- Identified the Maintenance lagging and leading KPIs.
- Identified the Maintenance assessment pillars.
- Identified the Maintenance assessment elements per pillar.
- Determined five maturity levels.
- Created maturity behaviors per maturity level per element.
- Reviewed the model to ensure alignment.
- Determined the criteria weighting per pillar.
- Determined the total scoring guidelines.
- Using the newly created model, Maintenance conducted a self-assessment.
- Using the newly created model, Maintenance from another business unit conducted the peer assessment.
- Based on the scoring outcome, Maintenance selected three areas for opportunity to focus on during the course of the year with the target to mature the scoring for the following year's assessment.
- Areas for opportunity were included within the owners annual performance review.

Ah Ha Moments:
Discovered there was a team member disconnect between how some KPIs were defined.

Lessons Learned:
In most cases a "one size fits all" model to be used across multiple sites does not work.

End Result:
Reduced Maintenance costs by ~$3M and improved equipment availability by 12% and reliability by 7% within a 4 year period.

4. Line & Point Kaizen
Use Case #1 (Application Benefits)
Step Reduction for a Machine Load Area

Business Case:
Reduce the number of employees assigned to loading a welder machine with metal shields. Currently, one unloads the shields in a warehouse due to cardboard dust not permitted in the process cleanroom environment and the other employee loads the shields into the machine. Target: Reduce headcount by one full-time employee (FTE) and reduce space by 400 sq. ft. for a total cost savings of $120,000 based on a 2-2-3 run schedule.

Process Applied:
Day 1:
- Conducted Genba Tour.
- Decided Target Area(s).
- Formed Team.
- Confirmed Team, Target Areas, and Target for Achievement.
- Team Completed a Kaizen A3 .
  - Team Leader
  - Team Co-Leader
  - Team Members
  - Description of the Business Case
  - Description of Current State / Problem to be Solved
  - Description of Future / Ideal State

Day 2:
- Presented the Kaizen A3 for Selection, Conducted Kaizen, and Identified Actions to Achieve Future / Ideal State (Kaizen A3 Detailed Implementation Plan).

Day 3:
- Kaizen Team Presented.
- Completed a Kaizen Sheet.
  - Before Kaizen
  - Problems
  - Activities / Actions
  - After Kaizen
  - The Results of Kaizen (Before to After)
- Conducted Confirmation Genba Tour.

Ah Ha Moments:
Learned that the vendor of the shields gained efficiency within their process by replacing the cardboard with plastic.

Lessons Learned:
Challenge everything while doing Kaizen.

End Result:
Cardboard replaced with plastic so one employee can unload and load the shields at the point of use. Cost savings at $120,000 (Note: the reduction of one headcount just eliminated one open job requisition and the 400 sq. ft. space was reallocated.)

Use Case #2 (Application Benefits)
Step Reduction Within a Warehouse Load Process

Business Case:
Improve Warehouse Load Area where currently three temporary employees are assigned to work across three stations in addition to one warehouse input step. Target: improve productivity, reduce one temporary employee headcount, reduce warehouse space and inventory for a total cost savings of $102,000.

Process Applied:
Day 1:
- Conducted Genba Tour.
- Decided Target Area(s).
- Formed Team.
- Confirmed Team, Target Areas, and Target for Achievement.
- Team Completed a Kaizen A3.
  - Team Leader
  - Team Co-Leader
  - Team Members
  - Description of the Business Case
  - Description of Current State / Problem to be Solved
  - Description of Future / Ideal State

Day 2:
- Presented the Kaizen A3 for Selection, Conducted Kaizen, and Identified Actions to Achieve Future / Ideal State (Kaizen A3 Detailed Implementation Plan).

Day 3:
- Kaizen Team Presented.
- Completed a Kaizen Sheet.
  - Before Kaizen
  - Problems
  - Activities / Actions
  - After Kaizen
- The Results of Kaizen (Before to After).
- Conducted Confirmation Genba Tour.

Ah Ha Moments:
Identified a safety / ergonomic constraint during the event.

Lessons Learned:
Plan into the Kaizen Event actions to prevent disruption to the process flow.

End Result:
Improved productivity by 31%, reduced one temporary employee headcount, reduced inventory by 200 pieces, and reduced floor space by 315 sq. ft. for a cost savings of $102,000.

USE CASES PHASE | PREDICTIVE

Reliability Centered Maintenance (RCM)

Use Case #1 (Application Benefits)
Predict Steel Mill Soaking Pit Wall Failure

Business Case:
Using data analytics and a PF-Curve (PF-Curves are generated from high risk items identified in Failure Mode & Effects Analysis (FMEA), determine the P-F Interval for Soaking Pit wall failure. Use this information to schedule wall repair during scheduled maintenance windows as compared to run to failure which results in loss of efficiency, scrap, loss of soaking pit availability, and slowing down or even idling the Melt Shop. Target: $300,000+ cost avoidance.

Process Applied:
- Collected historic data per soaking pit. Each soaking pit was treated independently of each other. Historic data included:

- Hours or cycles per soaking pit when wall failure occurred (run-to-failure event).
- Hours or cycles per soaking pit between when the wall failure was visually detected and when the wall failed completely.
- Hours or cycles per soaking pit from the last wall repair to visual detection of potential failure.
- Hours or cycles per soaking pit from last wall replacement to visual detection of potential failure.
- Created the PF-Curve by plugging in actual start of failure, potential failure (P), and functional failure (F) data.
- Used the PF-Curve to identify a time range where failure was likely to occur and based on that, aligned the scheduled corrective action with the next scheduled maintenance window.
- Established Condition-Based Maintenance (CBM) per soaking pit by leveraging the use of the PF-Curve and proactively inspecting the condition of the soaking pit walls. Notably, the soaking pits did not contain any sensors that detected wall deterioration.

Ah Ha Moments:
Only the potential point of failure can be measured in time with any certainty.

Lessons Learned:
- Ensure collected data are trusted and clean.
- If wall material is supplied by a different supplier from the supplier used with the original data collection, collect data for that supplier and complete another PF-Curve.
- Assign a P-value (probability of accuracy of the collected data) so that you can build a time buffer within the targeted time for corrective action.

End Result:
Cost avoidance of $350,000 annually

Use Case #2 (Application Benefits)
Implement Reliability Centered Maintenance in a Pilot Factory

Business Case:
Implement RCM in a selected pilot factory. Target: reduce the total cost of maintenance by $1.3M per year through the reduction of reactive and preventive maintenance by 50%, and increase predictive maintenance by >10% and proactive maintenance by >5%.

Process Applied:
- Provided RCM training to the Maintenance and Engineering teams.
- Provided Total Productive Maintenance (TPM) training to the Maintenance and Operation's leaders and designated TPM leaders.
- Created a Factory RCM A3 which included:
  - Team Members
  - Business Case
  - Current State
  - Target (Future) State
  - Detailed Implementation Plan
- Identified the infrastructure required, tools, and deliverables for the RCM deployment three phases:
  - Phase 1: Maintenance Organization Foundation
  - Phase 2: RCM Introduction / Engagement / Execution
  - Phase 3: RCM Sustainability
- Purchased a computerized maintenance management system (CMMS) for data collection, planning & scheduling, creating job plans and work orders, etc.
- Used the Factory RCM A3 Detailed Implementation Plan as a project management tool to track the status of all identified implementation actions required. Data regarding machine maintenance, for instance might include:
  - Quantity of units
  - Specification Make
  - Specification Model
  - Category
  - Manufacturer's Date
  - Date in Service
  - Age of the machine
  - Run Hours (Life)
  - Breakdown Hours
  - Risk of Impacting Output
  - Redundancy
  - Total Available Hours
  - Total Non Available Hours
  - Availability
  - Utilization
  - Reliability
  - Repair Spend
  - Preventive Maintenance (PM) Schedule
  - Predictive Maintenance (PdM) Schedule

- Part Obsolescence
- International Part Suppliers
- Domestic Part Suppliers
- Established PM and PdM schedules
- Established TPM schedule
- Created a A | B | C Parts Strategy
- Equipment sensor installment, where applicable

Ah Ha Moments:
Ensure the teams understand the difference between TPM and RCM.

Lessons Learned:
- Extracting data from the equipment is the most challenging step as older equipment you will have to manually collect the data whereas newer equipment captures the data within installed software programs.
- Get into the habit of reporting cost activity per maintenance category.

End Result:
Within a one year time period, achieved the reduction of reactive and preventive maintenance but fell short of the increase in predictive and proactive maintenance. However, the factory remained committed to continuing the focus on RCM and driving towards a more proactive and predictive maintenance culture. Actual Total Cost of Maintenance- $743,000

2. Simulations
Use Case #1 (Application Benefits)
Factory Startup with 3P

Business Case:
Validate the new factory layout (3D scale laid out on the factory floor) that was an end result of a 3P Event. Target: determine if adjustments / enhancements are required, thus increasing cost avoidance.

Process Applied:
- Ensured the 3D floor scale mock up was completed.
- Ran through several simulations of work flows and adjusted the mock up as required.
- Continued to run through several simulations of work flows with additional employees involved. Adjusted the mock up as required.

- Focused on identifying non-value add areas of the process and maximizing process flow.
- Focused on walking the process from supplier to delivery forward and backwards.
- Focused on walking through the process as if we were the product being assembled.
- Focused on walking through the process as if we were each individual part required to assemble the product.
- Transitioned the 3D Full-scale mockup into computer-aided design (CAD).
- Continued to run simulations and make the required adjustments. Standard work was created and updated to reflect all adjustments. Adjustments were made within CAD as well.
- Created material and information flow drawings to make visible and ensure they support the current state simulations.

- Maintained a Moonshine Equipment Sheet **(Figure 8-1)** where equipment and tools required were documented during the simulations.

MOONSHINE EQUIPMENT SHEET (TO BE USED DURING SIMULATIONS)		NO:	
		REQUESTEST DATE:	
REQUESTED BY:	SHIFT:	AREA/PROCESS NAME:	
Equipment Required (Sketch whenever applicable)			
Tools Required (Sketch whenever applicable)			
COMPLETED BY DATE:		REQUESTOR SIGN-OFF:	

Ah Ha Moments:
Even using pieces of candy to simulate flow through a factory layout is effective.

Lessons Learned:
- Layout will need to be re-evaluated / re-simulated anytime equipment / machine is replaced with one of a different size, thus changing the footprint.
- Make sure that the material and information flow supports the simulations.
- During the simulations, the team should ask, "Do we need this process; do we need to do this task(s)?"

End Result:
All risks associated with the new factory layout were identified during the simulation thus increasing cost avoidance.

Use Case #2 (Application Benefits)
Warehouse and Factory Consolidation

Business Case:
Consolidate a warehouse, located off site, within the manufacturing factory.
Target: Complete within 90 days at a total cost not to exceed $115,000

Process Applied:
- Ensured the warehouse footprint was fully and clearly taped off within the designated area within the factory.
- Ran simulations focused on:
  - Material & information flow
    - Fork truck routes
    - Part kit cart routes
- Aisle spacing for fork truck and kit cart travel.
- Ran additional simulations and made the appropriate adjustments to the layout.
- Updated the standard work.

Ah Ha Moments:
When relocating a warehouse, ensure the proper distance between the top of the racking and the fire suppression system.

Lessons Learned:

- Outsource the moving of the warehouse parts as compared to having company employees do it. This eliminates the financial risk of a company employee getting hurt or parts being damaged in transit.

End Result:

Completed the consolidation within the 90 days at the budget of $115,000.

ACRONYMS

INTRODUCTION
- BTOES: Business Transformation Operational Excellence Summit
- RCM: Reliability Centered Maintenance
- DNA: Deoxyribonucleic Acid
- S&P: Standards & Poor
- PDCA: Plan, Do, Check, Act

CHAPTER II: CHANGE COMMITMENT & ACCELERATION
- WIFM: What's in it For Me
- NASA: National Aeronautics and Space Administration

CHAPTER III: COMPETITIVE ADVANTAGE
- STE: Strategy to Execution
- CEO: Chief Executive Officer

CHAPTER IV: THE EXPERIENCE FACTOR
- QF: Quality Factor
- OEE: Overall Equipment Effectiveness

CHAPTER V: SIMPLIFY
- VOE: Voice of the Employee
- SEAL: Sea, Air, and Land
- 3P: Production Preparation Process
- SG&A: Selling, General, and Administrative
- FCF: Free Cash Flow
- CapEx: Capital Expense
- VSM: Value Stream Mapping
- DCO: Disruptive Cost Out
- VA: Value Add
- NVA: Non Value Add
- SMED: Single Minute Exchange of Dies
- BoB: Best of the Best
- WoW: Worst of the Worst
- FPY: First Pass Yield
- KPI: Key Performance Indicator
- RTY: Rolled Throughput Yield
- SPA: Single Point of Accountability
- MSDS: Material Safety Data Sheet

- BOM: Bill of Materials
- CTQ: Critical to Quality
- VOC: Voice of the Customer
- VA/VE: Value Add / Value Engineering
- DMB: Daily Management Board
- AR: Augmented Reality
- RYG: Red, Yellow, Green
- OPL: One Point Lesson

CHAPTER VI: CONTROL
- WIP: Work in Process
- TWI: Training Within Industry
- DMAIC: Define, Measure, Analyze, Improve, Control
- SCOT: Shipped Complete and On-Time
- SQDC: Safety, Quality, Delivery, Customer
- TRIR: Total Recordable Injury Rate
- DPU: Defects Per Unit
- IT: Information Technology

CHAPTER VII: PROACTIVE
- SWOT: Strengths, Weaknesses, Opportunities, Threats
- SDCA: Standardize, Do, Check, Act
- DMADV: Define, Measure, Analyze, Design, Validate
- IDDOV: Identify, Define, Develop, Optimize, Verify
- FTE: Full Time Employee
- JDI: Just Do It

CHAPTER VIII: PREDICTIVE
- METT-T: Mission, Enemy, Time, Troops, Terrain
- CMMS: Computerized Maintenance Management System
- PF: Potential Failure
- FMEA: Failure Mode & Effects Analysis
- CBM: Condition Based Maintenance
- TPM: Total Productive Maintenance
- PM: Preventive Maintenance
- PdM: Predictive Maintenance
- CAD: Computer Aided Design

CHAPTER IX: AN APPROACH TO INDUSTRY 4.0
- ROI: Return on Investment

FINAL THOUGHTS

My main objectives of writing this book was to provide the readers a proven model that will drive competitive advantage, provide a process to drive change commitment and acceleration, provide tool use cases, and tool caution points where companies can fail.

In business, I define insanity as choosing to remain in a reactive state while expecting to grow. Companies must choose to be proactive and leverage continuous improvement.

It is people who underpin the Model; without people, it is just words on paper. People must be enabled, encouraged, and supported to execute the model. This begins by developing a learning culture where people feel free to expand their thinking, be part of driving change, and be intense about company growth. I have been exposed to a lot of lip service over the years… if you seek organization growth, you must develop and grow your people.

Companies can avoid pitfalls such as a culture with a lack of trust by leveraging the intellect of their people as compared to trying to control them.

The level of descriptive detail that has been provided for each component of the 4 Phase Approach Model is sufficient enough to get a company started on the road to transformation or turnaround.

ALWAYS BE TRANSFORMING!

BIBLIOGRAPHY

1. Womack JP JD. Lean Thinking: Banish Waste and Create Wealth in Your Corporation. London, UK: Simon & Schuster;   2003.

2. The Global State of Operational Excellence: Critical Challenges & Future Trends. Business Transformation and Operational Excellence Summit. Research Report 2021/2022.

3. Lafley AG, Martin RL. Playing to Win: How Strategy Really Works. Boston. Harvard Business Review Press. 2013.

4. Porter ME. Competitive Advantage: Creating and Sustaining Superior Performance. New York: Free Press. 1985.

5. BrainyQuote.com. Jack Welch Quotes: BrainyMedia Inc; 2023 [Accessed July 3, 2023] Available from: https://www.brainyquote.com/quotes/jack_welch_173305.

6. goodreads.com. Leonardo da Vinci: Good Reads. [Accessed July 3, 2023]. Available from: https://www.goodreads.com/quotes/9010638-simplicity-is-the-ultimate-sophistication-when-once-you-have-tasted.

7. BrainyQuote.com. Albert Einstein Quotes: BrainyMedia Inc; 2023. [Accessed July 3, 2023]. Available from: https://www.brainyquote.com/quotes/albert_einstein_383803.

8. Peters T. PEP-20 - The Zen of Python. 2004 [Accessed July 3, 2023}. Available from: https://peps.python.org/pep-0020/.

9. Gryta T, Mann T. Lights Out: Pride, Delusions, and the Fall of General Electric. New York: First Mariner Books. 2020.

10. Womack JP, Jones DT. Lean Thinking: Banish Waste and Create Wealth in Your Corporation. London, UK: Simon & Schuster. 2003.

11. Lean Enterprise Institute. Value Steam Mapping. [cited July 3, 2023]. Available from: https://www.lean.org/lexicon-terms/value-stream-mapping/.

12. Phoenix Safety Management. Remember Charlie. 1994. Film. Running Time: 56 minutes.

13. Shingijutsu Global Consulting Inc. [cited July 3, 2023]. Available from: http://www.shingijutsu-global.com/en/index.html.

14. Liker J. The Toyota Way: 14 Management Principle from the World's Greatest Manufacturer. New York: McGraw-Hill. 2004.

15. AZ Quotes. Taiichi Ohno. [cited July 3, 2023]. Available from: https://www.azquotes.com/quote/1411459.

16. Leinwand P, Mainardi C, Kleiner A. Only 8% of Leaders Are Good at Both Strategy and Execution. Harvard Business Review. 2015;30

17. MIT Management Executive Education. Closing the Gap Between Strategy and Execution-course brochure. [Accessed July 3, 2023]. https://executive.mit.edu/course/closing-the-gap-between-strategy-and-execution/a056g00000URaZjAAL.html.

18. BrainyQuote.com. Alan Lakein Quotes: BrainyMedia Inc; [Accessed July 3, 2023.] Available from: https://www.brainyquote.com/quotes/alan_lakein_154655.

19. Lean Sensei International. [Accessed July 3, 2023]. Available from: https://www.leanagilesensei.com.

20. The Deming Institute. A Bad System Will Beat a Good Person Every Time. 2015 [Accessed July 3, 2023]. Available from: https://deming.org/a-bad-system-will-beat-a-good-person-every-time/].

21. Kanter RM. Ten Reasons People Resist Change: Which Ones are Hurting Your Company? Harvard Business Review. 2012(September).

Robert Cartia

Background

Rob has held leadership roles across multiple industries within Fortune 500 and private equity companies such as: Sony, PepsiCo, Allegheny Technologies Inc., General Electric, Johnson Controls International, Global Container Terminals, and REV Group. Over two decades of experience driving business operations and business transformation has resulted in over $1 billion dollars of cost savings and generated revenue. He is known for his work in Lean, Agile, Problem Solving, Six Sigma, Strategy Execution, Industry 4.0 / 5.0, Reliability Centered Maintenance, Change Acceleration & Commitment, Enterprise Business Systems and Operating Models. Rob is the Chairman of the Board at Parkview Community Federal Credit Union. In addition, he is a Maintenance Training Program Steering Committee member at Pittsburgh Institute of Aeronautics. Rob served in the United States Air Force where he functioned as a crew-chief (aircraft maintenance) assigned globally in high-pressure assignments.

A current C-level business transformation executive, Rob's primary industry background centers on technology and service-based environments. He has held positions within engineering, supply-chain, manufacturing, maintenance, quality, safety, and business operations / operational excellence. He was instrumental with the design, development, and implementation of new manufacturing lines and intelligent factory start-ups in both domestic and international environments within General Electric.

Experience

Rob's broad industry experience includes assignments in the Electronics,

Bottling, Metals, Rail & Mining, Energy, Automotive, Transportation & Logistics, Specialty Automotive, Aerospace & Defense, and Financial Services industries. He has delivered organizational value through an intense focus on Daily Management, Lean Management, and Breakthrough Management driven by Change Commitment & Acceleration within challenging environments.
One of the Six Sigma initiatives for which Rob was responsible for was a requisition to platform process which facilitates company new product launches. This initiative was a critical component of the company's process-improvement program and represented 1 of 40 big processes throughout a Fortune 5 company. Results included: 80% system reuse, 6-month reduction in cycle, and a 35% reduction in sole-source suppliers and overtime.

Key Skills
Rob focuses on accelerating organizational speed and efficiency via partnerships and key problem-solving capabilities to resolve critical business challenges. He functions as a business transformation leader and strategic business partner to organizational executive leadership teams across business value chains focused on the execution of business strategy.

Thought Leader
Rob is an energetic motivational speaker, accomplished communicator, and dedicated mentor.

Rob has made significant intellectual contributions in the areas of Performance Improvement, Strategy to Execution, Reliability Centered Maintenance, Supply Chain Optimization, Engineering & Product Development, Artificial Intelligence, Enterprise Business Systems and Operating Models. His ability to stay ahead of the latest trends in business and technology has Rob frequently asked to deliver presentations to organizations such as the Mississippi Enterprise for Technology at the NASA Stennis Space Center to include Stennis tenants such as National Oceanic and Atmospheric Administration and General Dynamics. He also presents annually at the Business Transformation and Operational Excellence Summit. Rob is a three-time published author and the developer of the Machine Order of Analysis©, Cause and Effect Check Sheet©, and the 4 Phase Approach to Competitive Advantage©.

Education / Certifications
Rob holds a MBA from the University of Pittsburgh and an Advanced Certificate for Executives (ACE) from MIT Sloan School of Management. In addition, Rob is a certified Lean-Six Sigma Master Black Belt and Quality Engineer.

www.ingramcontent.com/pod-product-compliance
Lightning Source LLC
Chambersburg PA
CBHW081816200326
41597CB00023B/4274